Clean and Renewable Energy

Clean and Renewable Energy

Edited by:

Jovan Pehcevski

ARCLER
PRESS

www.arclerpress.com

Clean and Renewable Energy

Jovan Pehcevski

Arcler Press
224 Shoreacres Road
Burlington, ON L7L 2H2
Canada
www.arclerpress.com
Email: orders@arclereducation.com

ISBN: 978-1-77407-780-1 (Hardcover)

DECLARATION

ABOUT THE EDITOR

Jovan obtained his PhD in Computer Science from RMIT University in Melbourne, Australia in 2007. His research interests include big data, business intelligence and predictive analytics, data and information science, information retrieval, XML, web services and service-oriented architectures, and relational and NoSQL database systems. He has published over 30 journal and conference papers and he also serves as a journal and conference reviewer. He is currently working as a Dean and Associate Professor at European University in Skopje, Macedonia.

TABLE OF CONTENTS

LIST OF CONTRIBUTORS

Guanjun Ding
Key Research Lab for Information, Beijing Information Technology Institute, Beijing, China

Bangkui Fan
Key Research Lab for Information, Beijing Information Technology Institute, Beijing, China

Teng Long
School of Information and Electronics, Beijing Institute of Technology, Beijing, China

Haibin Lan
Key Research Lab for Information, Beijing Information Technology Institute, Beijing, China

Yan Liu
School of Science, The Second Artillery Engineering University, Xi'an, China

Jing Wang
Key Research Lab for Information, Beijing Information Technology Institute, Beijing, China

Masataka Motoyama
Department of Aeronautics and Astronautics Engineering, Kyushu University, Fukuoka, Japan

Kenichiro Sugitani
Research Institute for Applied Mechanics, Kyushu University, Fukuoka, Japan

Yuji Ohy
Research Institute for Applied Mechanics, Kyushu University, Fukuoka, Japan

Takashi Karasudani
Research Institute for Applied Mechanics, Kyushu University, Fukuoka, Japan

Tomoyuki Nagai
Research Institute for Applied Mechanics, Kyushu University, Fukuoka, Japan

Shinsuke Okada
Department of Aeronautics and Astronautics Engineering, Kyushu University, Fukuoka, Japan

Ghassan Halasa
Electrical Engineering Department, University of Jordan, Amman, Jordan

Johnson A. Asumadu
Electrical and Computer Engineering Department, Western Michigan University, Kalamazoo, USA

Zhonglei Shao
Department of Electronic and Electrical Engineering, University of Strathclyde, Glasgow, UK.

Kwok Lun Lo
Department of Electronic and Electrical Engineering, University of Strathclyde, Glasgow, UK.

M. D. Berni
Interdisciplinary Center on Energy Planning (NIPE), State University of Campinas (UNICAMP), Campinas, Brazil

P. C. Manduca
Interdisciplinary Center on Energy Planning (NIPE), State University of Campinas (UNICAMP), Campinas, Brazil

S. V. Bajay
Interdisciplinary Center on Energy Planning (NIPE), State University of Campinas (UNICAMP), Campinas, Brazil

J. T. V. Pereira
Interdisciplinary Center on Energy Planning (NIPE), State University of Campinas (UNICAMP), Campinas, Brazil

J. T. Fantinelli
Interdisciplinary Center on Energy Planning (NIPE), State University of Campinas (UNICAMP), Campinas, Brazil

Rosario Miceli
Dipartimento di Ingegneria Elettrica, Elettronica delle Telecomunicazioni, di Tecnologie Chimiche, Automatica e Modelli Matematici, Università di Palermo, Palermo, Italy

Salvatore Favuzza
Dipartimento di Ingegneria Elettrica, Elettronica delle Telecomunicazioni, di Tecnologie Chimiche, Automatica e Modelli Matematici, Università di Palermo, Palermo, Italy

Fabio Genduso
Dipartimento di Ingegneria Elettrica, Elettronica delle Telecomunicazioni, di Tecnologie Chimiche, Automatica e Modelli Matematici, Università di Palermo, Palermo, Italy

Gillian Basso
IRTES-SET, UTBM, Belfort Cedex, France.

Nicolas Gaud
IRTES-SET, UTBM, Belfort Cedex, France.

Franck Gechter
IRTES-SET, UTBM, Belfort Cedex, France.

Vincent Hilaire
IRTES-SET, UTBM, Belfort Cedex, France.

Fabrice Lauri
IRTES-SET, UTBM, Belfort Cedex, France.

Hermes José Loschi
Department of Communications, Faculty of Electrical and Computer Engineering, University of Campinas (UNICAMP), Campinas/SP, Brazil

Julio Leon
Department of Communications, Faculty of Electrical and Computer Engineering, University of Campinas (UNICAMP), Campinas/SP, Brazil

Yuzo Iano
Department of Communications, Faculty of Electrical and Computer Engineering, University of Campinas (UNICAMP), Campinas/SP, Brazil

Ernesto Ruppert Filho
Department of Communications, Faculty of Electrical and Computer Engineering, University of Campinas (UNICAMP), Campinas/SP, Brazil

Fabrizzio Daibert Conte
Department of Communications, Faculty of Electrical and Computer Engineering, University of Campinas (UNICAMP), Campinas/SP, Brazil

Telmo Cardoso Lustosa
Department of Communications, Faculty of Electrical and Computer Engineering, University of Campinas (UNICAMP), Campinas/SP, Brazil

Priscila Oliveira Freitas
Department of Communications, Faculty of Electrical and Computer Engineering, University of Campinas (UNICAMP), Campinas/SP, Brazil

Morris Brenna
Politecnico di Milano – Department of Energy, Milan, Italy;

Ettore De Berardinis
CESI S.p.A., Milan, Italy;

Federica Foiadelli
Politecnico di Milano – Department of Energy, Milan, Italy;

Gianluca Sapienza
Politecnico di Milano – Department of Energy in Collaboration with ENEL Distribuzione S.p.A., Milan, Italy.

Dario Zaninelli
Politecnico di Milano – Department of Energy, Milan, Italy;

Zeyad A. Haidar
Electrical Engineering Department, King Saud University, Riyadh, Saudi Arabia.

Abdullah M. Al-Shaalan
Electrical Engineering Department, King Saud University, Riyadh, Saudi Arabia.

Mohamed Talaat
Electrical Power and Machines, Faculty of Engineering, Zagazig University, Zagazig, Egypt

Reda Edris
Electrical and Computer Engineering, Higher Technological Institute, 10th of Ramadan City, Egypt

Naglaa Ibrahim
Electrical and Computer Engineering, Higher Technological Institute, 10th of Ramadan City, Egypt

Fatma Omar
Electrical and Computer Engineering, Higher Technological Institute, 10th of Ramadan City, Egypt

Mohamed Ibrahim
Electrical and Computer Engineering, Higher Technological Institute, 10th of Ramadan City, Egypt

Nicole Griffin
UtiliWorks Consulting, Baton Rouge, LA, USA

Athens Gomes Silaban
UtiliWorks Consulting, Baton Rouge, LA, USA

Shi (Jessee) Zhang
Shenyang Agricultural University, Shenyang, China.
China University of Political Science and Law, Beijing, China.

Paula Costa
Laboratório Nacional de Energia e Geologia—LNEG, Lisboa, Portugal

Teresa Simões
Laboratório Nacional de Energia e Geologia—LNEG, Lisboa, Portugal

Ana Estanqueiro
Laboratório Nacional de Energia e Geologia—LNEG, Lisboa, Portugal

Vincent Nnaemeka Emodi
Department of Technology Management, Economics and Policy Program, College of Engineering, Seoul National University, Seoul, South Korea

Samson D. Yusuf
Department of Technology Management, Economics and Policy Program, College of Engineering, Seoul National University, Seoul, South Korea

Kyun-Jin Boo
Department of Technology Management, Economics and Policy Program, College of Engineering, Seoul National University, Seoul, South Korea

Félix Hernández
Institute of Economics, Geography and Demography (Spanish National Research Council), Madrid, Spain;

Miguel Hernández-Campos
Foundation Gómez Pardo, Madrid, Spain.

Hwa Meei Liou
Graduate Institute of Technology Management, National Taiwan University of Science and Technology, Taipei, Taiwan

Antonio Oscar Jr.
Federal University of Rio de Janeiro, Rio de Janeiro, Brazil

Wanderson Luiz Silva
Federal University of Rio de Janeiro, Rio de Janeiro, Brazil

Vera Ruffato
Federal University of Rio de Janeiro, Rio de Janeiro, Brazil

Renata Barreto
Federal University of Rio de Janeiro, Rio de Janeiro, Brazil

Marcos Freitas
Federal University of Rio de Janeiro, Rio de Janeiro, Brazil

LIST OF ABBREVIATIONS

ADEME	Agency for Environment and Energy Management
ACP	Alternative Compliance Payment
AVR	Automatic Voltage Regulator
AACC	Autonomic Communities
CHP	Combined heat and power
CAES	Compressed Air Energy Storage
CEF	Connecting Europe Facility
DCNS	Data Center Networks
DR	Demand response
DOD	depth of discharge
DLC	Direct Load Control
DNS	Direct Numerical Simulation
DER	Distributed energy resources
DER	Distributed Energy Resources
DG	Distributed Generation
EES	Electric Energy Storage
EPRI	Electric Power Research Institute
ECN	Energy Commission of Nigeria
EAM	Enterprise Agent Manager
EDSOSG	European Distribution Systems Operators Association for Smart Grids
EEGI	European Electricity Grids Initiative
EERA	European Energy Research Alliance
EII	European Industrial Initiatives
EHV	Extra high voltage
FERC	Federal Energy Regulatory Commission
FPC	Federal Power Commission
FIT	Feed-in tariff
FDM	Finite Difference Method

GCC	Generator Control Centre
GRTU	Generator Remote Terminal Unit
GIS	Geographical Information System
GWP	Global Warming Potential
GHG	Greenhouse gases
HV	High voltage
HWSPS	Hybrid wind/solar power system
IPP	Independent power producers
ISO	Independent systems operator
IEC	International Electrochemical Commission
ISO	International Organization for Standardization
ISPs	Internet Service Providers
ITC	Investment Tax Credit
JMD	Jordan Meteorological Department
LCE	Levelized cost of energy
LPSP	Loss of Power Supply Probability
LV	Low Voltage
MPPT	Maximum power point tracker
MVDNs	Medium-Voltage Distribution Networks
METI	Ministry of Economy, Trade and Industry
MAS	Multi-Agent Systems
NCERD	National Center for Energy Research and Development
NCEP	National Center for Environmental Predictions
NSO	Nigeria Standards Organization
NMESs	Numerical Models of the Earth System
OLTC	On-Load Tap Changer
PMSG	Permanent magnetic synchronous generator
PMV	Predicted Mean Vote
PDF	Probably density function
PDCs	Project Development Committees
QoS	Quality of Service
RTU	Remote terminal units
RE	Renewable energies
REIP	Renewable Energy Incentive Program

RET	Renewable energy technologies
RPS	Renewable portfolio standard
SHP	Small hydropower
SDNO	Smart Distribution Network Operation
SRECs	Solar Renewable Energy Credits
SNCE	Spanish National Commission on Energy
SON	Standards Organization of Nigeria
SOC	State of charge
SHV	Super high Voltage
SCADA	Supervisory Control and Data Acquisition
SDM	Supply and demand matching
TC	Technical Committees
TOU	Time of use
VPP	Virtual Power Plant
VTO	Voltage Threshold Overall

PREFACE

Energy is the ability to do, move and work. We live in a world of energy and virtually everything around us is based on the use of some kind of energy. So far known natural processes and phenomena can be explained by several forms of energy: kinetic energy, potential energy, thermal energy, gravity, elasticity, electromagnetism, chemical energy, nuclear energy and mass.

The main sources of energy that can be found in nature are: Sun energy (photosynthesis), evaporation, water and air flow, Earth energy (geothermal and gravity energy), and tide, the consequence of the gravitational force of the Sun, Moon and Earth.

When it comes to natural and primary forms of energy, we can divide them into renewable and non-renewable ones, given the time potential for their depletion. Renewable means that these sources owe the fact that energy is consumed at an amount not exceeding the rate at which it is produced in nature. Renewable energy reserves are often estimated for exploitation over several million years. This makes a fundamental difference to non-renewable resources, whose reserves have been estimated to last several tens or hundreds of years and have lasted incomparably longer.

Renewable energy can be divided into two main categories:

- Traditional renewable energy sources, such as biomass and energy from large hydropower plants, and
- New renewable energy sources, such as solar, wind, geothermal and the like.

The development of renewable energy sources especially from wind, water, sun and biomass is important for several reasons:

- These energy sources play a very important role in reducing carbon dioxide (CO_2) emissions into the atmosphere. This is also an important part of the USA, EU policies etc.
- Increasing the share of renewable energy sources increases the energy sustainability of a country's system. At the same time, it helps to improve the security of energy delivery and thus reduces dependence on imports of energy raw materials as well as electricity.

- It is expected in the near future that renewables will become economically competitive with conventional energy sources.

This edition covers different topics from clean and renewable energy, including solar and wind energy, smart grids, analysis of renewable energy, and case studies from different countries.

Section 1 focuses on solar and wind energy, describing technical properties of wind and solar photovoltaic power; improving the power generation performance of a solar tower using thermal updraft wind; wind-solar hybrid electrical power production to support national grid: a Jordan case study; and overview of research on optimization of integrated solar/wind power generation systems.

Section 2 focuses on smart grids, describing energy efficiency and renewable energy technologies using smart grids; a perspective on the future of smart grid distribution; a framework for qualifying and evaluating smart grids approaches with focus on multi-agent technologies; a prospective study on energy management systems and energy efficiency in smart grids; and a voltage control in smart grids based on sensitivity theory.

Section 3 focuses on analysis of renewable energy, describing reliability evaluation of renewable energy share in power systems; a new approach for converting renewable energy to stable energy; the utility perspective of powering renewable programs; and technical analysis and enlightenment of renewable energy.

Section 4 focuses on case studies from different countries, describing a GIS methodology for planning sustainable renewable energy deployment in Portugal; the necessity of the development of standards for renewable energy technologies in Nigeria; the development of the renewable energy technologies in Spain; the development of electricity grid, smart grid and renewable energy in Taiwan; and a case study of biofuels and solar energy that evaluates the renewable energy vulnerability to climate change in Brazil.

SECTION 1:
SOLAR AND WIND ENERGY

1

Analysis of Technical Properties of Wind and Solar Photovoltaic Power

Guanjun Ding[1], Bangkui Fan[1], Teng Long[2], Haibin Lan[1], Yan Liu[3], Jing Wang[1]

[1]Key Research Lab for Information, Beijing Information Technology Institute, Beijing, China
[2]School of Information and Electronics, Beijing Institute of Technology, Beijing, China
[3]School of Science, The Second Artillery Engineering University, Xi'an, China

ABSTRACT

Most of electricity power in China comes from coal and hydropower. Already, China must import nearly half of its oil. Concerns about power capacity shortages and air pollution are all adding urgency and pressure to switch to alternative technologies and renewable energy. Among renewable energy sources, wind power and solar photovoltaic power are playing key roles in China, and are the fastest-growing power generation technologies. So this paper focuses on them and analyzes the corresponding technical properties of them. First of all, wind power transforms the kinetic energy from the wind into electricity by using wind turbines. Thus the basic components of wind turbines are described. Wind speed is an important factor to wind

Citation: Ding, G., Fan, B., Long, T., Lan, H., Liu, Y. and Wang, J. (2013), "Analysis of Technical Properties of Wind and Solar Photovoltaic Power". Journal of Applied Mathematics and Physics, 1, 39-44. doi: 10.4236/jamp.2013.14008.

energy. So the features of wind speed are analyzed, and the wind energy is calculated. Second, the technical properties of solar photovoltaic power are discussed, including photovoltaic cells and modules, battery, inverter and photovoltaic controller. Photovoltaic energy is also analyzed and calculated. Third, the environmental impacts of wind power and solar photovoltaic power are presented. Finally, the relevant conclusions are drawn.

Keywords: Wind Power; Solar Photovoltaic Power; Technical Properties; Environmental Impacts

INTRODUCTION

With the serious problem of environmental pollution caused by fossil fuel, wind and solar photovoltaic power have emerged as two of the most attractive clean technologies and they are planned to be major sources of future electricity needs [1]. Wind and solar photovoltaic power have gained extensive interests, and they are the most mature and cost effective resources among different renewable energy technologies [2]. Wind power is a source of renewable power which comes from air current flowing across the earth's surface.

Solar photovoltaic power comes from the sunlight irradiates the earth's surface. They are affordable and sustainable. Since wind and sunlight are free and inexhaustible, the price of wind and solar photovoltaic power is stable, which is unlike electricity from fossil fuel powered source that depends on fuel whose price is costly and may vary considerably. Nevertheless, the major challenge of wind and solar photovoltaic power is that they are the intermittent power supplies, because wind doesn't always blow and sometimes there is no sunlight. Figure 1 shows the intermittent nature of wind power. Due to the intermittency characteristics of wind and solar photovoltaic power, it results in the variability, unpredictability, and uncertainty of wind and solar sources [3].

Thus the integration of wind and solar facilities to utility grid presents a major challenge to power system operator, it would cause some problems when integrating a large amount of wind and solar photovoltaic power into the existing electrical grid. Such integration has significant impact on the optimum power flow, transmission congestion, power quality issues, system stability, load patch, etc. [4,5]. So, in order to analyze the impacts on power system plan and operation in depth, the technical properties of wind and solar photovoltaic power must be considered. This is the aim of this paper. The paper analyzes and studies the corresponding technical properties, including wind turbine, wind speed, wind power and energy, photovoltaic cells and modules, battery and inverter, photovoltaic controller and energy, to provide the analysis basis and reference for further rescarch of power system containing wind and solar power.

TECHNICAL PROPERTIES OF WIND POWER

Wind Turbine

A wind turbine is the machine which converts the kinetic energy from wind into mechanical energy [6], as shown in Figure 2. The mechanical energy is then converted to electricity. The modern wind turbine is a sophisticated piece of machinery with aerodynamically designed rotor and efficient power generation, transmission and regulation components [7,8]. Its size ranges from a few Watts to several Million Watts. Most of today's commercial machines are horizontal axis wind turbine with three bladed rotors. The followings introduce briefly some of the main components of wind turbine.

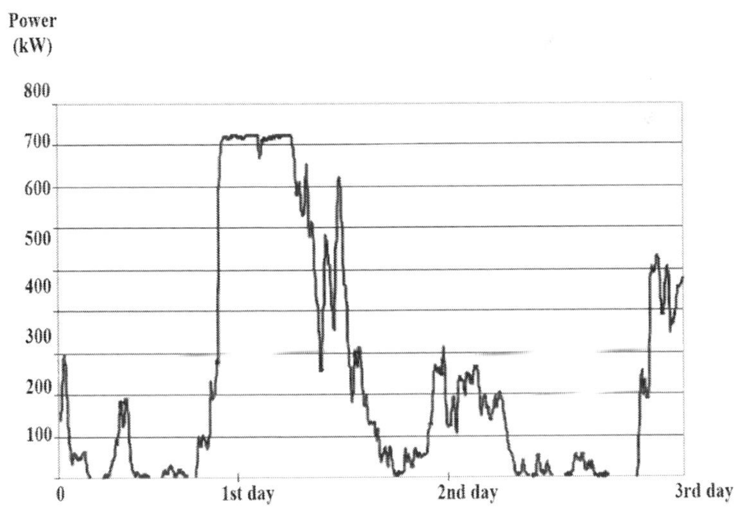

Figure 1. Diagram of the intermittent nature of wind power

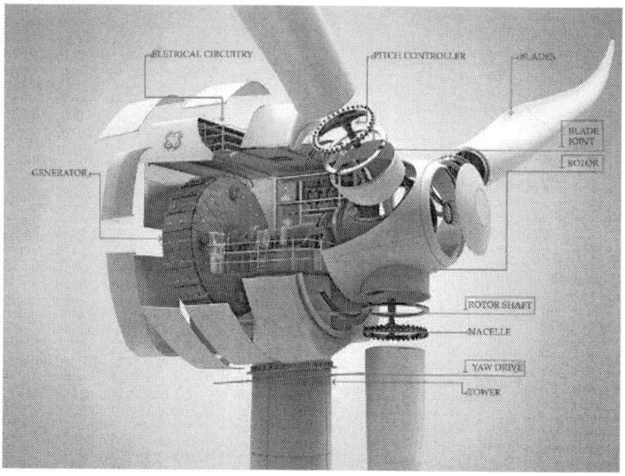

Figure 2. Schematic diagram of a wind turbine.

- ***Rotor/Blades***: The blades together with the hub are called the rotor. The rotor drives the generator by harnessing the kinetic energy in the wind. The blades are aerodynamically shaped to best capture the wind. The amount of energy wind turbine can capture is proportional to the rotor sweep area.
- ***Generator/Alternator***: It is the part which produces electricity from the kinetic energy captured by the rotor. A generator produces DC power and an alternator produces AC power, depending on the application for the turbine.
- ***Nacelle***: It is the housing which protects the essential motorized parts of a turbine.
- ***Gearbox***: It is used for most turbines above 10kW to match the rotor speed to the generator speed.

The power curve of a wind turbine is a graph that represents the turbine output power at different wind speed [9]. It is usually provided by the turbine's manufacture. Figure 3 shows an example of a wind turbine power curve. It is notable that the output power is zero at speed from 0 to 2.5 m/s. This happens due to there is not sufficient kinetic energy in the wind to move the wind turbine rotor. Normally the manufactures provide technical data sheets where the start up wind speed of the turbine is given. Generally lower start up wind speeds result in higher energy coming from the turbine. Sometimes the power curve information may be shown in a table format. Some manufactures offer the exact power values at different wind speed and present this in a table. The power curve is then obtained by plotting the table values.

Wind Speed

No other factor is more important to the amount of wind power available to a wind turbine than wind speed. Because the power in wind is cubic function of wind speed, changes in speed produce a remarkable effect on power. Doubling the wind speed does not double the power available it increases a whopping eight times. Wind speed varies over time. Wind speed varies by the minute, hour, day, season, and even by year. It is influenced by weather system, the local land terrain and its height above the ground surface.

The average speed is composed of winds above and below the average. The cube of the average wind speed is always less than the average of the cube of wind speed. Using the average annual wind speed alone in the power equation would not give the right results. It's the wind speed above the average that contributes most of the power. Since wind speed varies, it is necessary to capture this variation in the model used to predict the energy production. It is usually done using probability functions to describe wind speed over a period of time. The variation in wind speed is best described by a probably density function (PDF) [10], as shown in Equation (1). A PDF is used to model the wind velocity variation. It provides the probability that an event occurs between two end points. The area under the curve between any two speeds greater than zero will equal the probability that wind will blow somewhere between those two speeds.

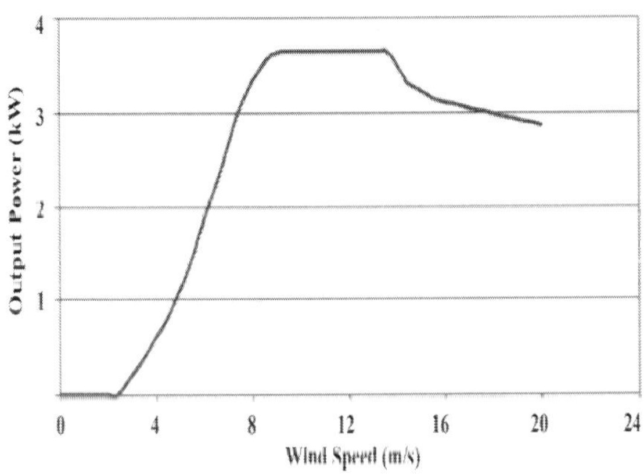

Figure 3. Schematic diagram of the power curve of a wind turbine.

$$f(v)=\frac{\alpha}{\theta}\left(\frac{v}{\theta}\right)^{\alpha-1}e^{-\left(\frac{v}{\theta}\right)^{\alpha}} \tag{1}$$

where v represents in this case the wind speed, α is the shape factor and θ is the scale factor. For a given average wind speed, a smaller shape factor indicates a relatively wide distribution of wind speeds around the average, while a larger shaper factor indicates a relatively narrow distribution of wind speeds around the average.

Wind speed calculated includes two types, i.e., the arithmetic mean wind speed and the cubic root cube wind speed [11].

The arithmetic mean wind speed is what normally known as the average wind speed. It is given by:

$$v_{ave}=\int_{v_{min}}^{v_{max}} f(v)\cdot v\cdot dv \tag{2}$$

where f (v) is the Weibull PDF, v is the data vector of measured wind speed, v_{min} is the minimum measured wind speed and v_{max} is the maximum measured wind speed.

The use of arithmetic mean wind speed tends to underestimate the electric power production. The cubic root cube wind speed produces a better estimate of actual power production. To find the cubic root cube average speed, the data vector of wind speed is elevated to the cube and multiplied by the PDF. The function is integrated between v_{min} and v_{max}. Then it is elevated to cubic root. The result is the cubic root cube average speed, which is defined as:

$$v_{crc}=\sqrt[3]{\int_{v_{min}}^{v_{max}} f(v)\cdot v^{3}\cdot dv} \tag{3}$$

Likewise, where f (v) is the Weibull PDF, v is the data vector of measured wind speed, v_{min} is the minimum measured wind speed and v_{max} is the maximum measured wind speed.

Wind Power and Energy

Wind power is a function of air density, the area intercepting the wind and the wind speed [12]. It is calculated as below:

$$P = \frac{1}{2} \varepsilon \cdot S \cdot v^3 \qquad (4)$$

where P is the output power in watts, ε is the air density in kg/m^3 , S is the area in m^2 where wind is passing and v is the wind speed in m/s.

Wind energy is power over some unit of time. The energy production can be calculated substituting the average wind speed value in power Equation (4). Then multiplying the Equation (4) by the hours of the period, the energy is obtained as shown below:

$$E = \frac{1}{2} \varepsilon \cdot S \cdot v^3 \cdot H \cdot D \qquad (5)$$

where E is the total energy in Wh, H is hours, D is days, and the variable v can be either the arithmetic mean wind speed or the cubic root cube wind speed, but using the cubic root cube wind speed is better estimation of the average wind speed than the arithmetic mean wind speed.

TECHNICAL PROPERTIES OF SOLAR PHOTOVOLTAIC POWER

Photovoltaic Cells and Modules

Photovoltaic cells consist of semiconductor material, i.e., silicon, which is at present the most often utilized [13]. Photovoltaic cells have electric fields which can force electrons to flow in a certain direction. The flowing of electrons is a current and it can be used externally. Because of depending on the behavior of the solar resource, the electricity produced by photovoltaic cells is intermittent. The technology used is modular, since it could be connected to pre-existing installations of photovoltaic panel, as shown in Figure 4, and replaced individually. The efficiency of photovoltaic cells decreases with increases in temperature. Photovoltaic panel reacts directly to sunlight. The chances of change in weather could block sunlight, such as clouds, rain and sandstorms. Photovoltaic modules are made up of interconnected photovoltaic cells. The cells in the modules are connected together in a designed configuration to deliver useful current and voltage. The cells connected in parallel increase the current output, while the cells connected in series increase the voltage output. Groups of several photovoltaic modules connected together are called solar array.

Battery

Battery is a device which stores DC electrical energy in electrochemical form for later use [14]. Due to not all batteries rechargeable, they are divided in two categories. The first category can't be recharged and only converts chemical energy to electrical energy. The second can be recharged. Because energy is lost in the chemical reaction during charging or recharging, the conversion efficiency of battery is not perfect. The internal components include positive and negative electrodes plates [15]. The life of battery is directly related to how deep the battery is cycled. Discharge depth refers to how much capacity could be used from the battery. Most systems are designed for regular discharges up to 40 - 80 percent.

Figure 4. Schematic diagram of photovoltaic panel.

Temperature can affect the performance of battery. The capacity of battery will increase at higher temperature and decrease at lower temperature. The life of battery will increase at lower temperature and decrease at higher temperature.

Equation (6) describes how to calculate the number of batteries connected in series to reach the voltage required by the system. Where V_{DC} is the DC system voltage (Volt), V_B is the battery voltage (Volt).

$$N_S = \frac{V_{DC}}{V_B} \quad (6)$$

Equation (7) describes how to calculate the number of batteries connected in parallel to reach the Amp hours required by the system. Where C_R is the required battery bank capacity (Ah), C_S is the capacity of selected battery (Ah).

$$N_P = \frac{C_R}{C_S} \quad (7)$$

The total number of batteries needed can be obtained by multiplying the number in series and the number in parallel as shown in equation (8).

$$N = N_S \cdot N_P \quad (8)$$

Inverter

The inverter converts the DC electrical energy to AC electrical energy, which can then be used to operate AC devices like the ones plugged in to most household electrical outlets [16]. The normal output AC waveform of inverter is sine wave with frequency of 50/60 Hz. Inverters are available including three different categories, i.e., grid tied battery less, grid tied with battery back-up and stand alone. The most popular inverters are grid tied battery less. These inverters directly connect to the public utility, using the utility power as storage battery. When the sun shines, the electricity comes from the photovoltaic via the inverter. If the photovoltaic array produces more power than used, the excess is sold to utility power company through the electric meter. Inverter sizing contains calculating the number of inverters needed for the photovoltaic system. When selecting an inverter must have a DC voltage equal to inverter DC voltage and have an AC voltage and frequency equal to home and utility values.

$$N_I = \frac{P_L}{P_I} \quad (9)$$

Equation (9) describes how to calculate the number of inverters needed. Where P_L represents the maximum continuous power loading home consumes, P_I is the maximum power supplied by the inverter.

Photovoltaic Controller

The photovoltaic controller operates as a voltage regulator. Its primary function is to prevent the battery from overcharged. When the batteries are fully charged, the controller will stop or decrease the amount of current flowing into the battery. The average efficiency of the controller ranges from 95% to 98%. If high current is required, two or more controllers can be used. When more than one controller used, it is necessary to divide the array into sub-arrays.

Each subarray is wired into the same battery bank. The photovoltaic controller contains the five different types, i.e., shunt controller, single stage series controller, diversion controller, pulse width modulation controller and maximum power point tracking controller. The maximum power point tracking controller is the best for photovoltaic system at present. It allows the controller to track the maximum power point of the array throughout the day to delivery the maximum available solar energy to system. Before maximum power point tracking controller was available, the array voltage would be pulled down slightly above the battery voltage while charging. When selecting a controller must be sure → ensured it has an output voltage rating equal to the nominal battery voltage, also the maximum photovoltaic voltage should be less than the maximum controller voltage rating.

Photovoltaic

Energy Hourly average solar radiation values are usually used to calculate the photovoltaic energy (kWh) generated for one year at a specific site, as shown in Equation (10).

$$E_Y = P_{PM}\left(R_{AVE}\right)\cdot T_{PM}\left(R_{AVE}\right)\cdot 365 \tag{10}$$

where E_Y is the yearly expected photovoltaic energy production at a given site (kWh), P_{PM} (R_{AVE}) is the photovoltaic module output power at an average hourly solar irradiation (R_{AVE}), T_{PM} (R_{AVE}) is the time of the sun hit the photovoltaic module at RAVE, the product of 365 is to change daily to yearly quantities.

Environmental Impacts of Wind and Solar Photovoltaic Power

The impacts of wind power on environment are relatively small. The impacts on wildlife in operation stage are related to the noise of wind turbines. Due to the power lines associated with wind farms, it may cause electromagnetic radiation or possible forest fire. For offshore wind farms, the impacts are those with regard to fishing, navigation and effects on marine life. The power lines buried under the seabed could have an impact on breakable ecosystems. If the offshore wind farms close to shore, they may have an impact on birds.

The first aspect of the environmental impacts of solar photovoltaic power is about aesthetics when its components installed. Photovoltaic panels may occupy some spaces, e.g., building roofs, road and railroad margins. The second aspect of the impacts is associated with the photovoltaic cells production. During the production process and the arrangement stage, it uses some poisonous materials, which are harmful to environment.

CONCLUSIONS

Wind energy transforms the kinetic energy from the wind into usable electricity by utilizing wind turbine. Wind turbine is composed basically of a tower base, three blades and a generator at the middle hub where the motion of the blades is transformed into electricity by means of inductance. The advantage of power curve is that it includes the wind turbine efficiency. Wind speed is a quite important element to wind energy. By the Weibull probably density function (PDF), the wind velocity variation can be described accurately. Combined with the Weibull PDF, the arithmetic mean wind speed and the cubic root cube wind speed can be derived and calculated. Based on the obtained wind speed, wind power and energy can be calculated. Solar photovoltaic power generates electricity from the sunlight radiation. When sunlight strikes photovoltaic cells, the direct current is generated. Photovoltaic modules consist of interconnected photovoltaic cells. To increase the voltage and current output, photovoltaic cells are connected in series and parallel respectively. For storing DC electrical energy for later use, battery is necessary.

The features of battery could be influenced by temperature. The life of battery is directly relevant to how deep it is cycled. To convert DC electrical energy to AC electrical energy for AC devices running, the inverter must be provided. The type of grid tied battery is less commonly used in some inverters types. The photovoltaic controller serves as a role of voltage regulator to prevent battery from overcharged. To calculate the photovoltaic energy at a given site, hourly average solar radiation values are used. From these derived technical properties, the analysis basis and reference can be provided for further study on the power system including wind and solar photovoltaic power.

ACKNOWLEDGEMENTS

This work was financially supported by the 52nd General Program of China Postdoctoral Science Foundation (2012M521837)

REFERENCES

1. G. Jose, "The Case for Renewable Energies," Thematic Background Paper at the International Conference for Renewable Energies, Bonn, Germany, 2004. http://www.renewables2004.de
2. J. Martin, "Learning in Renewable Energy Technology Development," Ph.D. Thesis, Utrecht University, Utrecht, 2005.
3. M. Patel, "Wind and Solar Power Systems," 2nd Edition, Taylor & Francis, 2006.
4. M. S. Lu, C. L. Chang and W. J. Lee, "Impact of Wind Generation on a Transmission System," Proceedings of Power Symposium, NAPS, 2007.
5. C. I. Chai, W. J. Lee, P. Fuangfoo, M. Williams and J. Liao, "System Impact Study for the Interconnection of Wind Generation and Utility System," Proccedings of IEEE I&CPS Conference, Clearwater Beach, Florida, 2005.
6. J. F. Manwell, J. G. Mcgowan and A. L. Rogers, "Wind Energy Explained," Wiley Press, New York, 2002. http://dx.doi.org/10.1002/0470846127
7. F. Bianchi, H. D. Battista and R. Mantz, "Wind Turbine Control Systems," Springer-Verlag Press, London, UK, 2007.
8. K. Tan and S. Islam, "Optimum Control Strategies in Energy Conversion of pmsg Wind Turbine System without Mechanical Sensors," IEEE Transactions on Energy Conversion, Vol. 19, No. 2, 2004, pp. 392-399. http://dx.doi.org/10.1109/TEC.2004.827038
9. G. Ramtharan, N. Jenkins and L. Anaya, "Modelling and Control of Synchronous Generators for Wide Range Variable Speed Wind Turbines," Wind Energy, Vol. 10, No. 3, 2007, pp. 231-246. http://dx.doi.org/10.1002/we.219
10. E. G. Pavia and J. J. Brien, "Weibull Statistics of Wind Speed over the Ocean," Journal of Climate and Applied Meteorology, Vol. 25, No. 10, 1986, pp. 1324-1332. http://dx.doi.org/10.1175/1520-0450(1986)0252.0.CO;2
11. G. J. Herbert, S. Iniyan, E. Sreevalsan and S. Rajapandian, "A Review of Wind Energy Technologies," Renewable and Sustainable Energy Reviews, Vol. 11, No. 6, 2007, pp. 1117-1145. http://dx.doi.org/10.1016/j.rser.2005.08.004
12. The Wind Indicator, "Wind Energy Facts and Figures from Wind

Power Monthly," Windpower Monthly News Magazine, Denmark, USA, 2005.

13. B. S. Borowy and Z. M. Salameh, "Optimum Photovoltaic Array Size for a Hybrid Wind/PV System," IEEE Transactions on Energy Conversion, Vol. 9, No. 3, 2004, pp. 482-488. http://dx.doi.org/10.1109/60.326466
14. B. S. Borowy and Z. M. Salameh, "Methodology for Optimally Sizing the Combination of a Battery Bank and PV array in a Wind/PV Hybrid System," IEEE Transactions on Energy Conversion, Vol. 11, No. 2, 2006, pp. 367-375. http://dx.doi.org/10.1109/60.507648
15. A. Smeets, "Investigation of the Solar Production of Silicon Nitride by Carbothermic Reduction of Silicon Dioxide," Diplomarbeit ETH-Swiss Federal Institute of Technology, Swiss, 2003.
16. PVDI, "Solar Energy International (Photovoltaic Design and Installation Manual)," New Society Press, New York, 2007.

2

Improving the Power Generation Performance of a Solar Tower Using Thermal Updraft Wind

Masataka Motoyama[1], Kenichiro Sugitani[2], Yuji Ohya[2], Takashi Karasudani[2], Tomoyuki Nagai[2], Shinsuke Okada[1]

[1]Department of Aeronautics and Astronautics Engineering, Kyushu University, Fukuoka, Japan
[2]Research Institute for Applied Mechanics, Kyushu University, Fukuoka, Japan

ABSTRACT

The purpose of this study is to improve the efficiency of the power generation system of a solar tower using fluid dynamics. The power generation system of a solar tower can be designed and constructed at relatively low cost. However, the energy output tends to be low for its physical size compared with other renewable energy production systems. The technical and scientific improve- ment of these types of generation systems has lost its momentum since the shutdown of the well- known Spanish pilot plant "Manzanares Solar Chimney" in 1989, although it still has the potential to play a role in renewable energy in the future. We have focused on the tower component of the system to seek possible enhancements of the power output of the internal

Citation: Motoyama, M., Sugitani, K., Ohya, Y., Karasudani, T., Nagai, T. and Okada, S. (2014), "Improving the Power Generation Performance of a Solar Tower Using Thermal Updraft Wind". Energy and Power Engineering, 6, 362-370. doi: 10.4236/epe.2014.611031.

turbine. As a result of our fluid dynamic shape optimization, a diffuser-shaped tower was employed to increase the internal flow speed of a scaled model. The results show a remarkable improvement in the power output of the internal wind turbine.

Keywords: Solar Tower, Thermal Updraft, Wind Turbine, Diffuser

INTRODUCTION

A solar tower [1] that can generate electricity in a simple structure, and enables easy and less costly maintenance, has considerable advantages. In particular, a tower of the thermal updraft wind type has significant potential as a renewable energy power plant. One of the few problems for this type is the low power electric generation capacity per unit area. Consequently, the solar tower needs to be enlarged to achieve an adequate performance. If we can improve its performance in power generation by some method, the solar tower may become competitive when compared with other renewable energy systems. In this study, we show an example of optimizing the tower shape in terms of its fluid dynamic characteristics to raise its efficiency in generating power.

Power plants that use towers either have a thermal updraft as for the solar tower shown in Figure 1 [2] - [4] or are of another type of tower called an energy tower. For the solar tower the sun's energy is collected by a heat collector below the tower creating a thermal updraft that passes into the transparent collector and rotates the wind turbine. The power plant mechanism in the energy tower involves spraying the upper part of the tower with water, creating a circulating and increasing downdraft flow within the tower that rotates the wind turbine [5] [6] . The solar tower, which we focus on in this study, is based on the simple structure conducted in the demonstration experiment in Manzanares, Spain, from 1982 to 1989 by the German engineer, J. Schlaich. The concept of the solar tower is not new. Leonardo da Vinci designed a chicken barbecue with a windmill driven by the hot upwind in the chimney [2] . In developing the original idea for the purpose of exploiting solar energy, simple techniques—in this case, the greenhouse, the chimney and the windmill—were combined in an uncomplicated collector and conversion system. Thus the solar tower was born.

However, a thermal updraft wind type of solar tower power plant using thermal updraft wind is, for example, in Manzanares, where the height is 194 m, and the collector is 46,000 m^2, a very large scale plant, but generates little power for its size, and remains at 44.19 MWh per annum. On the other hand, for example, if we spread solar panels over the same area of land, 8.5 GWh of power is generated (assumed capacity factor 13%) under conditions of Japan's greatest annual sunshine (1420 kWh/m^2).

Since the pilot plant in Manzanares was completed, further development worldwide has been minimal and any scientific technical research has hardly been carried out, although the Menzanares solar tower still has the potential to play a role in renewable energy in the future. Recently, with priorities such as non-nuclear power generation and CO_2 reduction, renewable energy sources have been attracting attention in the search for energy security. Consequently, other renewable energy sources such as wind-generated electricity, where there is adequate wind, and solar cells, where there is adequate sunshine, are generally recognized as the most promising options. Fundamental experiments on how much power is generated depending on the size of the basic component of the solar tower, have not been performed (power generation expectation and the optimization study [7] [8]). We applied a diffuser shape as the first step in optimizing the tower shape in a trial to increase power generation using fluid dynamics. We performed an experiment to compare the updraft wind velocity flowing through a diffuser-shaped tower with a typical cylindrical form. This experiment was conducted with two large scale models and one small scale model to confirm that the results complied with the scaling law. It demonstrated that when upsizing the scaling law remained valid.

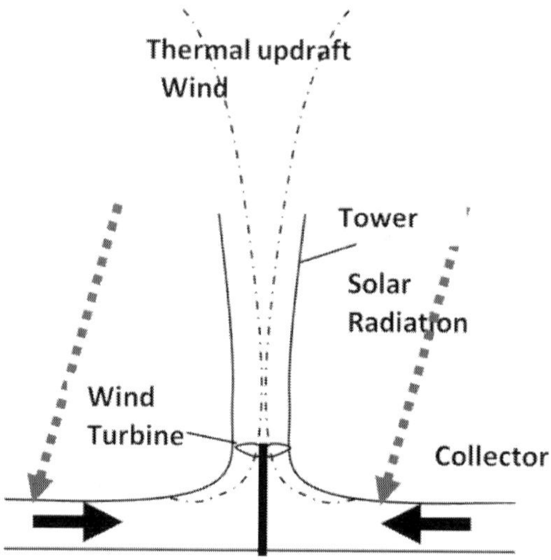

Figure 1.Mechanism of solar tower.

PREVIOUS EXPERIMENTS

Preliminary experiments were conducted on mini-models of solar towers [9] [10] . The mini-models consisted of cylindrical and diffuser models with a 0.66 m radius collector and a tower 0.4 m high. If the external flow around the solar tower is assumed to be negligible, the dominant flow is from heat convection. The flow field is approximated by natural convection from the temperature difference between the flow temperature heated by the solar thermal energy and the ambient temperature. In this laboratory experiment, the flow field was the natural convection driven by the temperature difference between the flow temperature Θ2 from the controlled floor surface temperature at the bottom of the tower, and the ambient temperature Θ1. The ambient temperature was defined as the temperature surrounding the exit of the tower Θ1. The temperature difference was $\Delta\Theta = \Theta 2 - \Theta 1$. The flow created by the sun and ambient temperature was simulated as shown above. Figure 2 shows the experimental results from measuring the velocity of the updraft. In the cylindrical tower, the velocity at 70 mm above the heated floor was 0.55 m/s. In the diffuser tower the velocity at the same point was faster at 0.77 m/s. When the temperature difference ΔΘ was approximately 20°K, 70 mm above the floor, the velocity of the updraft in the diffuser tower exceeded that in the cylindrical tower. If the rotor of a wind turbine is set at this point, a larger power output is expected.

PAST NUMERICAL SIMULATION

A computational fluid dynamics analysis was conducted by simulating the experimental model. The state variables, which were the velocity components, pressure and temperature, were computed using Direct Numerical Simulation (DNS) based on the Finite Difference Method (FDM). The time integration algorithm was the variant of the fractional step.

Method [11] [12] : The pressure was determined by computation of the Poisson equation, solved using the Successive Over Relaxation (SOR) method. A second-order central scheme was used to discretize the spatial derivatives except for the convective terms, which were discretized using the modified third-order upwind differencing scheme based on an interpolation method [13] [14] . Figure 3 shows the updraft flow velocity augmentation ratio, which was the updraft flow velocity of the diffuser tower normalized against the updraft flow velocity of the cylindrical tower. The straight line is Least Square Estimate. The augmentation ratios of the numerical and experimental values were almost the same. Even the numerical values showed a maximum augmentation ratio 70 - 90 mm from the floor, which gave the augmentation effect from the diffuser tower.

EXPERIMENT

The experiment with a larger model was conducted in a room. Figure 4 shows the experimental model. The towers used were a cylindrical type and a diffuser type with an open angle of 8° (4° each side). The towers were 2 m high.

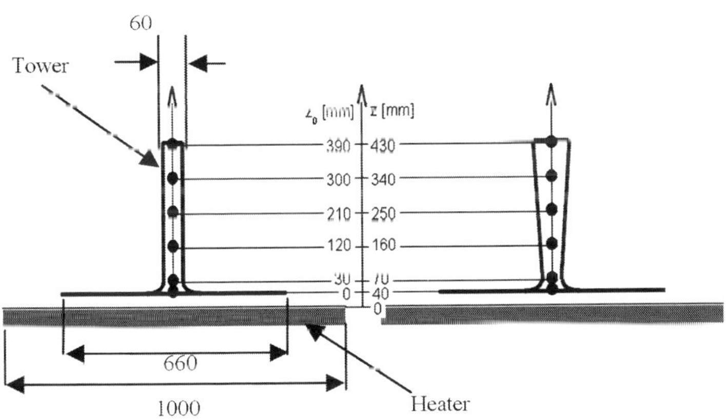

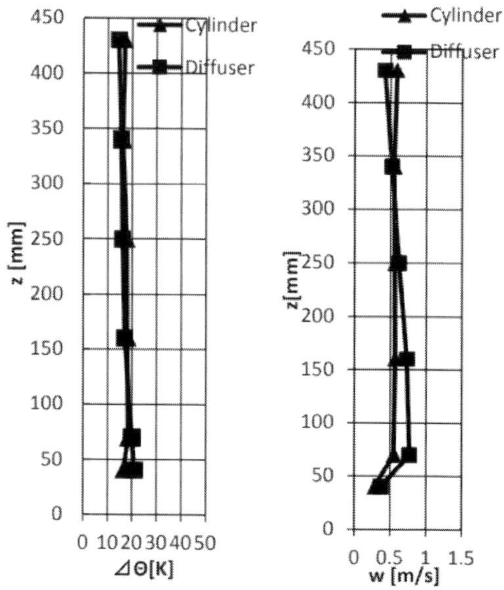

Figure 2. Measuring point of cylindrical tower and diffuser tower, right hand side temperature difference and updraft wind velocity along the vertical center line.

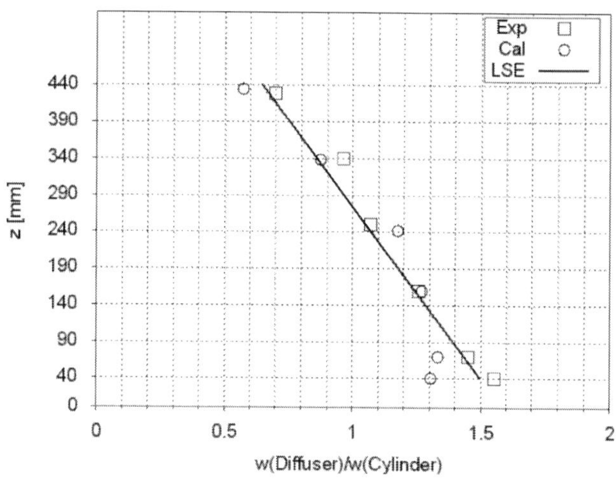

Figure 3. Updraft wind velocity distribution of laboratory experiment and computational fluid dynamics (Temperature Difference 30°C).

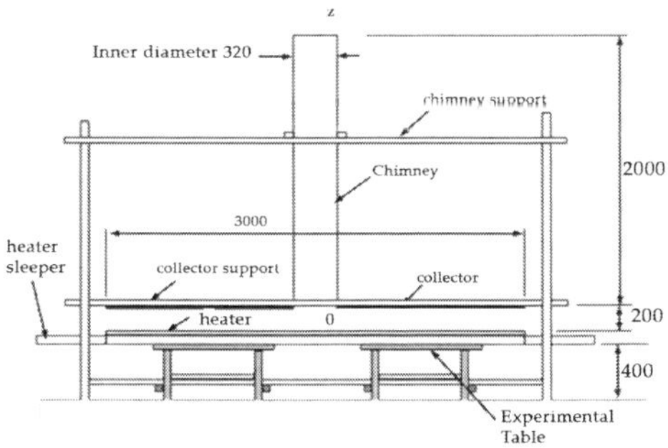

Figure 4. Experimental model and the configuration.

To exchange the towers, and make it easy to insert or remove the wind turbine, the tower part was easily disassembled. One side of the collector section was 3 m and it was supported by spacers 20 cm from the floor. The collector was made of acrylic sheet. It was divided into three sections. To ensure the thermal updraft wind did not leak from the gaps between the divided collectors, they were covered with a vinyl sheet. The panel unit was 1.5 m × 1.5 m 9 mm thick iron plate with 8 × 375 W heater elements. The heating elements were connected to the four panel unit. A temperature sensor (Pt100) was attached in the center of a unit panel. Using the signals from these sensors, a temperature controller controlled the temperature of the panel unit. In the experiment, the temperatures of the iron plates were controlled by the controller and heaters to maintain the panel units at the required temperatures. The updraft wind velocity was measured by inserting an Ultra Sonic Anemometer into the tower from the top to 5 cm above the blade of the wind turbine. The signal from the Ultra Sonic Anemometer was fed to a PC and monitored using an FFT Analyzer. Figure 5 shows the distribution of flow velocity. The location of the Ultra Sonic Anemometer was 6 to 7 cm between the center and edge of the inside of the tower. Note, as is shown in Figure 5, the distribution of wind velocity was almost uniform, and it was assumed to be the same over the horizontal plane inside the tower.

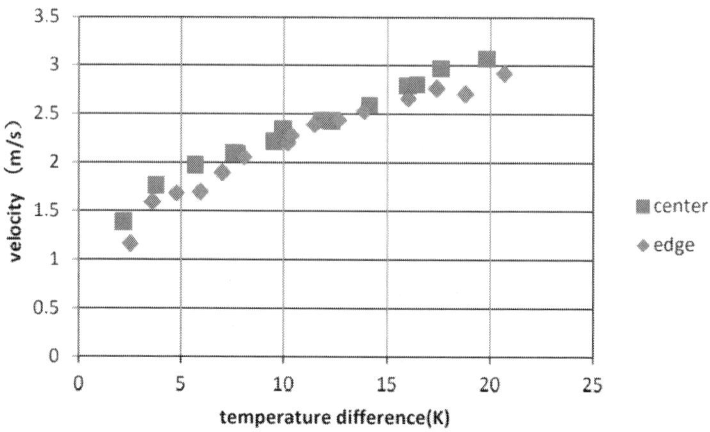

Figure 5. Velocity measured inside the tower.

RESULTS AND DISCUSSION

Without Wind Turbine

The dimensionless parameters which are dominant in the flow field of thermal convection are the Reynolds number Re and Richardson number Ri. When these two dimensionless parameters are equal, dynamic similarity is maintained. It is difficult for Ri and Re to be equal simultaneously. Re had the greatest difference between the real scale and experimental models, but at the tower throat section(radius 0.32 m), the updraft wind was 2.5 m/s, and Re ? 52000. This Reynolds number seemed to be close to real scale flow. On the other hand, Ri and the wind velocity had the following relationship between the real scale and experimental models:

$$Ri = \frac{gL}{W^2} \cdot \frac{\Delta\Theta}{\Theta} = \frac{gl}{w^2} \cdot \frac{\Delta\theta}{\theta}, \tag{1}$$

$$W = w\sqrt{\frac{L}{l}\frac{\Delta\Theta/\Theta}{\Delta\theta/\theta}}, \tag{2}$$

where, L, W, Θ are the reference length, reference velocity and reference temperature for each, capitalized parameters are for the real scale, and lower case parameters, l, w, θ are for the experimental model. Equation (2) is derived from Equation (1) and provides the transform for the expression between the real scale and experimental model. Initially, as stated in Section 3, we compared the result for the wind velocity of a small scale mini-model [9] [10] with that of the experimental model. If l is for the mini-model in the preliminary experiment and L is the present experiment, the scale is l:L = 1:5. Under natural convection, if the Ri of both the experimental and preliminary experimental models is the same, the velocity for the experimental model is evaluated by the root of the scale ratio multiplied by the velocity for the preliminary model.

Figure 6 shows, on the left, a two-way arrow, which is the scaled-up difference of the diffuser tower, and on the right, a two-way arrow, which is the scaled-up difference of the cylindrical tower. From these results, we can draw conclusions: The ratio of the velocity between mini-model and larger model is 2.2 - 2.5 and the values are almost equal to $\sqrt{5}$ i.e. from the scaling effect. We confirmed the wind velocity increased in proportion to the root of the scale ratio. Figure 6 also shows two measurements of the wind velocity inside the diffuser and cylindrical towers at this time. From this result, the ratio between the wind velocity inside the diffuser and cylindrical towers was almost 1.5 - 1.8 times for a temperature difference $\Delta\Theta$ 20 - 35°K at this time.

Case When the Wind Turbine Is Present

Test of Power Generation for the Wind Turbine

The test of power generation was conducted to check the characteristic of the wind turbine for experimental use. Figure 7 shows an experimental model. To investigate the optimal tip speed ratio λ and power coefficient Cp for this small wind turbine (rotor radius 0.15 m), Cp is given by Equation (3) [15] :

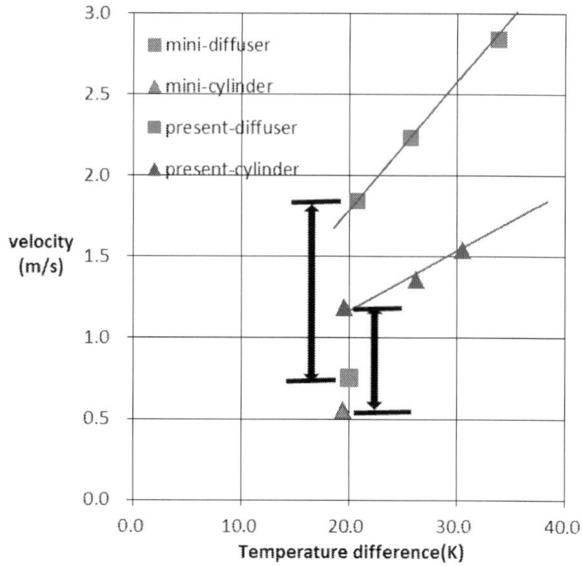

Figure 6. Comparison of wind velocity between mini-model (l = 40 mm high) and experimental model (L = 2000 mm high).

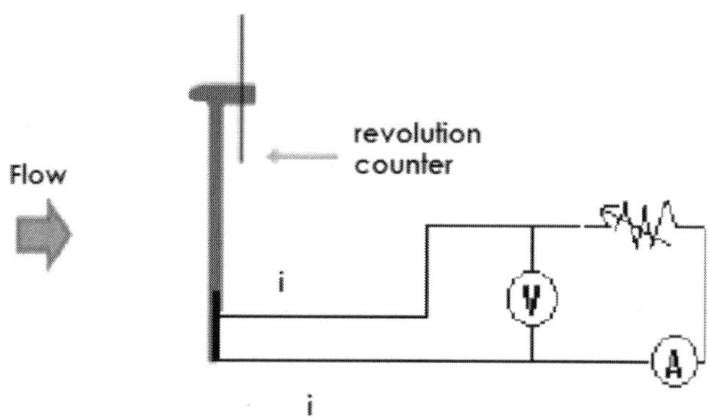

Figure 7. Model of wind turbine.

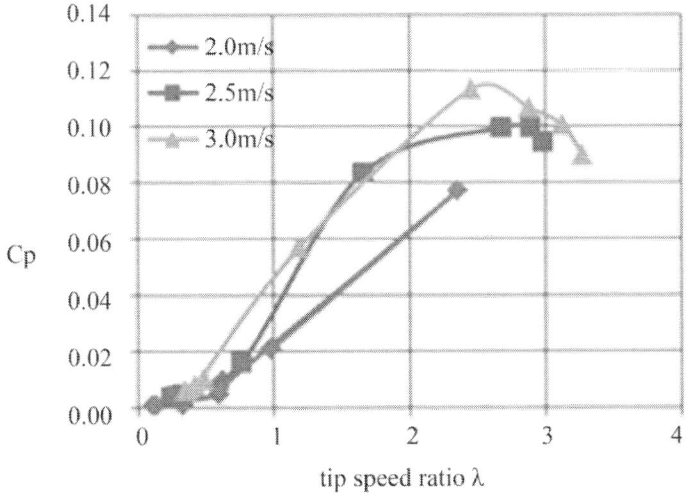

Figure 8. Cp vs. tip speed ratio λ.

$$Cp = \frac{P}{\frac{1}{2}\rho A U^3}. \quad (3)$$

To find out the optimal tip speed ratio λ_w the variable resistance is adjusted as in Figure 7. Figure 8 shows the results of the performance test. In the case of a wind velocity of 2.5 - 3.0 m/s as shown in Figure 8, Cp is the maximum at around a tip speed ratio of $\lambda_w = 2.5 - 2.9$.

Influence of the Wind Turbine on the Velocity

Figure 9 shows the relationship between the flow temperature and vertical velocity at 0.05 m above the rotor when it is rotating. Installing the wind turbine creates a resistance. The wind velocity decreases in comparison with the case with no wind turbine. In addition, the vertical wind velocity inside the diffuser tower is 1.7 times greater than that inside the cylindrical tower, i.e. (1) By installing the wind turbine, 1.9 m/s wind velocity (temperature difference 20°K, diffuser type) was slowed down to 1.7 - 1.8 m/s (Figure 9) whereas the velocity was slightly slowed down to 1.3 m/s in the case with no wind turbine (temperature difference 38°K, cylindrical). (2) Even if a wind turbine was installed, the ratio between the velocity of the diffuser and cylindrical towers was about 1.7, which is almost the same ratio with or without the wind turbine. As a consequence, the velocities for the diffuser and cylindrical towers are clearly different.

Performance of Power Generation

Next we discuss the power. Figure 10 shows the turbine rotation speed for the diffuser tower was greater than for the cylindrical tower. We measured the speed of rotation of the rotor with a revolution counter. The increasing speed of rotation of the rotor relates directly to the power output. Here, we define power as P, Torque as T_r and number of rotation ω, the relationship is given by Equation (4):

$$P = T_r \cdot \omega . \quad (4)$$

The power output is expected to increase for the diffuser tower. Figure 11 shows the power output vs. temperature difference.

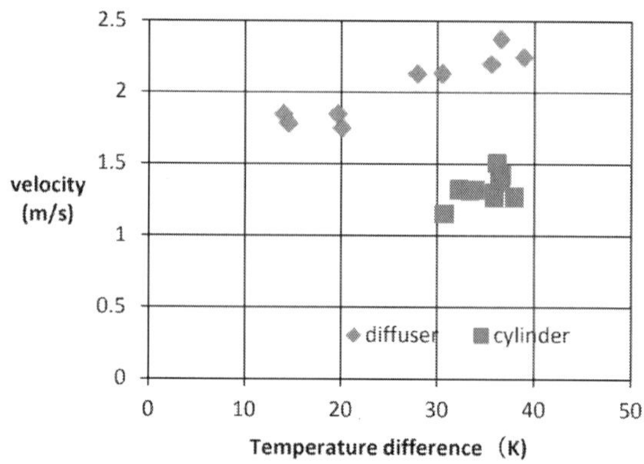

Figure 9. Temperature vs. velocity.

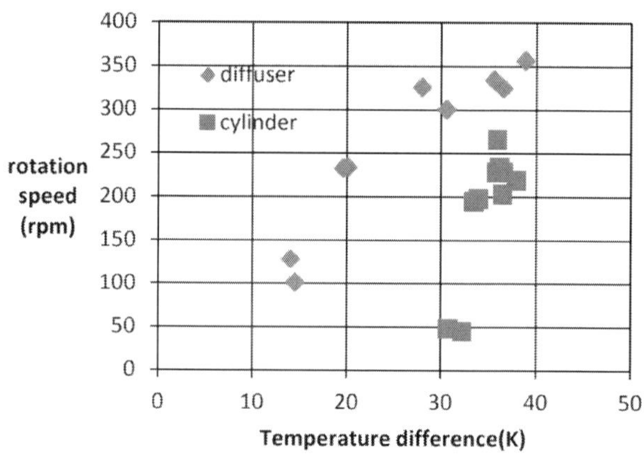

Figure 10. Temperature difference vs. rotational speed.

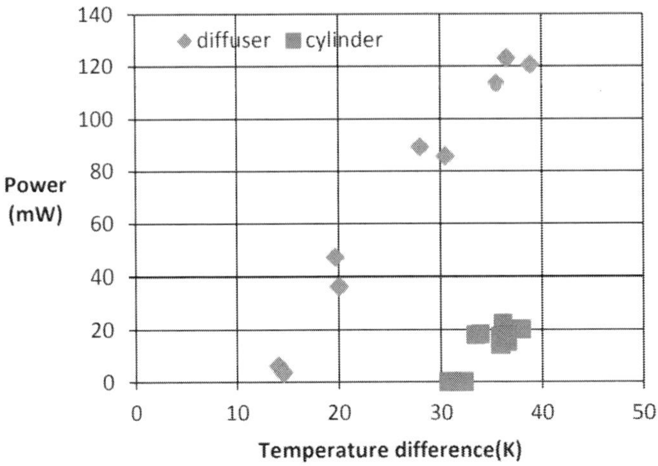

Figure 11. Temperature difference vs. power.

From the results in Figure 11, whereas the ratio of velocity between the diffuser and cylindrical towers is 1.7, the power output for the diffuser tower turns out to be 6 times greater than for the cylindrical tower. This is because the power output is proportional to the cube of the velocity.

In addition, from the relationship of the output for the diffuser vs. flow temperature, the diffuser output produces power from a lower temperature difference than the cylindrical output. In using the diffuser shape tower for a solar tower, the wind velocity exceeds that for a cylindrical type tower.

Furthermore, even if a wind turbine is installed, the ratio of the speed increase is constant, the speed of rotation of the rotor increases and the power output increases considerably.

CONCLUSIONS

The purpose of this research is to focus on the shape of the solar tower, and increase the power output by changing the structure from the conventional cylindrical tower to a diffuser type tower, thus increasing the power output.

As a result of previous preliminary experiments conducted by Okada, the mini-model of a diffuser type tower achieved greater wind velocity than a cylindrical tower. In these experiments, we made the size of the model 5 times larger than the mini model. By installing a wind turbine, we actually measured the obtained power output and verified it.

1) Initially, similar to the mini-model, we measured the wind velocity of the diffuser and cylindrical towers without a wind turbine. As the result, the wind velocity for the diffuser tower was greater than for the cylindrical tower by a factor of 1.5 - 1.8 times. The velocity of this model was almost $\sqrt{5}$ times greater than for the mini-model. A scaling law was established for the solar tower.
2) Next, by installing a wind turbine, the change in velocity and power output was measured. The ratio of velocity between the diffuser and cylindrical towers was almost 1.7, similar to the case for the temperature difference ?Θ without a wind turbine. Because the power output was proportional to the cube of the velocity, the power output obtained in the diffuser tower was 6 times greater than that for the cylindrical tower.

From these results, we concluded that the diffuser tower, which has the advantage of a larger power output than the cylindrical tower, was the preferred option for a solar tower.

ACKNOWLEDGEMENTS

Masataka Motoyama, Kenichiro Sugitani, Yuji Ohya, Takashi Karasudani, Tomoyuki Nagai, Shinsuke Okada (No. 2424161).

REFERENCES

1. Schlaich, J. (1996) The Solar Chimney. Axel Menges, Germany.
2. Haaf, W., Friedrich, K., Mayr, G. and Schlaich, J. (1983) Solar Chimneys—Part I: Principle and Construction of the Pilot Plant in Manzanares. International Journal of Solar Energy, 2, 3-20. http://dx.doi.org/10.1080/01425918308909911
3. Haaf, W. (1984) Solar Chimneys—Part II: Preliminary Test Results from the Manzanares Pilot Plant. International Journal of Solar Energy, 2, 141-161. http://dx.doi.org/10.1080/01425918408909921
4. Schlaich, J., Bergermann, R., Schiel, W. and Weinrebe, G. (2005) Design of Commercial Solar Updraft Tower Systems—Utilization of Solar Induced Convective Flows for Power Generation. Journal of Solar Energy Engineering, 127, 117-124. http://dx.doi.org/10.1115/1.1823493
5. Altman, T., Carmel, Y., Guetta, R., Zaslavsky, D. and Doytsher, Y. (2005) Assessment of an "Energy Tower" Potential in Australia Using a Mathematical Model and GIS. Solar Energy, 78, 799-808. http://dx.doi.org/10.1016/j.solener.2004.08.025
6. Omer, E., Guetta, R., Ioslovich, I. and Gutman, P.O. (2008) Optimal Design of an "Energy Tower" Power Plant. IEEE Transactions on Energy Conversion, 23, 215-225. http://dx.doi.org/10.1109/TEC.2007.905349
7. Gannon, A.J. and von Backström, T.W. (2000) Solar Chimney Cycle Analysis with System Loss and Solar Collector Performance. Journal of Solar Energy Engineering, 122, 133-137. http://dx.doi.org/10.1115/1.1314379
8. Koonsrisuk, A. and Chitsomboon, T. (2007) Dynamic Similarity in Solar Chimney Modeling. Solar Energy, 81, 1439- 1446. http://dx.doi.org/10.1016/j.solener.2007.03.003
9. Okada, S., Ohya, Y., Uchida, T. and Karasudani, T. (2006) Wind Convergence Performance Evaluation of VT- WCONVS. Proceedings of 19th Wind Engineering Symposium, Tokyo, 29 November 2006, 145-150. (in Japanese)
10. Okada, S., Ohya, Y., Uchida, T. and Karasudani, T. (2006) Convergence of Wind Using VT-WCONVS (Vertical Type Wind Convergence Structure). Proceedings of Renewable Energy, Yokohama, 2006.
11. Ohya, Y. and Uchida, T. (2008) Laboratory and Numerical Studies of the Atmospheric Stable Boundary Layers. Journal of Wind Engineering

and Industrial Aerodynamics, 96, 2150-2160. [Citation Time(s):1]

12. Kim, J. and Moin, P. (1985) Application of a Fractional-Step Method to Incompressible Navier-Stokes Equations. Journal of Computational Physics, 59, 308-323. http://dx.doi.org/10.1016/0021-9991(85)90148-2
13. Kawamura, T., Takami, H. and Kuwahara, K. (1986) Computation of High Reynolds Number Flow around a Circular Cylinder with Surface Roughness. Fluid Dynamics Research, 1, 145-162. http://dx.doi.org/10.1016/0169-5983(86)90014-6
14. Kajishima, T. (1994) Upstream-Shifted Interpolation Method for Numerical Simulation of Incompressible Flows. Transactions of the Japan Society of Mechanical Engineers B, 60, 3319-3326. (in Japanese)
15. Manwell, J., Mcgowan, J. and Rogers, A. (2009) Wind Energy Explained. Wiley, New York.

3

Wind-Solar Hybrid Electrical Power Production to Support National Grid: Case Study - Jordan

Ghassan Halasa[1], Johnson A. Asumadu[2]

[1]Electrical Engineering Department, University of Jordan, Amman, Jordan
[2]Electrical and Computer Engineering Department, Western Michigan University, Kalamazoo, USA

ABSTRACT

The paper presents the next generation of power energy systems using solar- and wind-energy systems for the country of Jordan. Presently with the oil prices are on the rise, the cost of electrical power production is very high. The opportunity of a large wind and solar hybrid power production is being explored. Sights are chosen to produce electricity using the wind in the Mountains in Northern Jordan and the sun in the Eastern Desert. It is found that the cost of windmill farm to produce 100 - 150 MW is US$290 million while solar power station to produce 100 MW costs US$560 million. The electrical power costs US$0.02/kWh for the wind power and US$0.077 for the solar power. The feasibility for using wind and solar energies is now when the price oil reaches US$ 100.00 per barrel. The paper also

Citation: G. Halasa and J. Asumadu, "Wind-Solar Hybrid Electrical Power Production to Support National Grid: Case Study - Jordan," Energy and Power Engineering, Vol. 1 No. 2, 2009, pp. 72-80. doi: 10.4236/epe.2009.12011.

discusses different power electronics circuits and control methods to link the renewable energy to the national grid. This paper also looks at some of the modern power electronics converters and electrical generators, which have improved significantly solar and wind energy technologies.

Keywords: solar energy, wind energy, hybrid energy system

INTRODUCTION

The concept of solar and wind energies dates back to nearly 7,000 years ago [1]. However, in the late 1800s the Danes developed the first wind turbines to produce commercial electricity [1–4]. In the early 1900s smallscale wind turbines became more widely used around Europe especially in the rural areas for producing electricity using old car generators and carved rotors. The wind power brought electricity to the rural areas and the electrical power was used to charge batteries to run radios and to draw water from deep wells [2]. Except in Denmark where wind power production and research continued, wind power did not play any major role in the generation of electricity until the late 1900s.

The rapid growth of solar and wind powers is due in part to favorable global political climate towards these energies, efforts to reduce carbon dioxide (CO_2) and greenhouse gases (GHG) and other power plant pollutants, global awareness of climate changes, and the urgency to develop renewable energy sources. Other factors such as lucrative tax incentives and legislation mandating national renewable energy standards have accelerated the march towards solar and wind energies. For example in the US, some states have enacted "renewable portfolio standard (RPS)" law that requires utilities to sell a certain percentage of the energy from sustainable energy sources within reasonable stipulated times. Even though Europe and North America have the largest installed capacity of wind turbine capacity, China, India, and developing world have the biggest potential for wind power [5].

This paper examines the capacity and potential for electricity-generating solar- and wind-turbines installed In the Eastern and Northern part of Jordan. The Jordan Meteorological Department (JMD) has histological data on wind speeds and sunshine days in areas of the country that can be used to assess the potential for solar and wind energies, and other applications. Jordan has excellent sunshine covering more than 80% of the country (on the average of 330-day in a year). The average wind speed in Jordan is 7 m/s (at 10 meters height above any obstacles within 100 meters) in some parts of the country. Presently, the total renewable energy power generation capacity is about 1% of power generation in Jordan. It is expected that the share of renewable energy in electrical power production will be 15% in the future. Wind Solar alternatives are essential for growth, finance, and the political environment. The cost of wind power has reduced from the cost of power production from US$ 0.09.5 per kilowatt-hour to less than US$0.02 for wind energy production and to US$0.076 cents for solar power production. This is very significant because developing countries, which depend on external sources to finance major energy projects, may be able to finance small scale solar and wind energies projects from their own resources and faster. In this paper the electrical and power calculations for solar and wind utilization to support the national grid in Jordan will be analyzed. This paper also looks at some of the modern power electronics converters and electrical machines, which have improved significantly solar and wind energy technologies to make them acceptable and embraced as cost effective and renewable energy.

THE EXISTING JORDAN'S NATIONAL GRID

Jordan is interconnected in one national grid. The grid covers most of the populated areas of the country from Aqaba, on the far south to Irbid in the far north. Overhead transmission line link Syria in the north, Palestine in the west, while undersea cable links Egypt in the south. Future countries to be connected to the Jordan national grid include Lebanon, Iraq and Turkey.

The major generation centers are the Aqaba Thermal Power Station in the far south, Al-Hussein Thermal Power Station in Zarka near Amman, and Al-Risha near the Iraqi border. The Aqaba Power Station uses gas supplied through pipelines from Egypt. The pipeline extends to Amman. Future expansion of this gas line is expected to go to Syria and eventually Turkey. Al-Hussein power station uses fuel oil imported from Iraq. Al-Risha power station uses locally produced gas. There are several small units scattered in different districts belonging to older utilities. These units are used during peak demands. The oldest and the highest power production plant in Jordan is the Al-Hussein Thermal Power Station. It is the most expensive because it uses imported oil and also uses air-cooling systems, that consume quite amount of energy, to cool the turbines. A small pilot plant uses biogas produced by sewerage treatment plant.

Another pilot plant uses wind energy near the sight proposed in North Jordan. Aqaba Power Station uses Egyptian gas supplied by gas pipeline. This pipeline already extended to Amman. Future expansion of this gas line is expected to go to Syria, and eventually Turkey. Al-Hussein power station uses fuel oil imported fro Iraq. Al-Risha power station uses locally produced gas. There are several small units scattered in different districts belonging to older utilities. These units are used in peak demands.

CONVENTIONAL ELECTRICAL PRODUCTION COST

The Kingdom of Jordan is considered an emerging country in the Middle East; it has almost no natural resources. The country imports most of its oil needs from neighboring countries at market prices. Oil and gas imports are huge burden on the country's national economy. Electricity is generated by burning imported gas and oil, limited generation from hydro, windmills, and biogas. When oil prices rose to extremely high levels last summer, Jordanians experienced continuous increases in electricity prices. It is now urgent and essential to deploy other alternatives for electrical generation, which is the use of solar and wind energy for electrical generation.

As shown in Table 1, Jordan in 2007 produced a total of 13,001 GWh of electrical energy and consumed 10,553 GWh. The average per capita electricity consumption in Jordan in 2007 was 2277 kWh as compared to 2075 kWh in 2006, resulting in annual growth rate of 9.7%. Table 1 shows generating capacity and electrical energy production by type of generation for 2007. The state-owned utility National Electric Power Company (NEPCO) currently carries out almost all electricity production in Jordan. Al-Hussein Power Plant (with capacity of 400 MW) and the Aqaba Power Plant (with capacity of 650 MW) are the country's two main power generation facilities. Jordan has modest reserves of natural gas of 230 billion cubic feet and has developed one gas field at Al-Risha in the eastern desert near the Iraqi border. The current output of this field is around 30 million cubic feet per day. Al-Risha field is used to fuel one nearby power plant, which generates about 10% of Jordan's electricity. For several years, Jordan has been exploring the option of importing natural gas from Egypt. In 1999, a decision was made to delay imports until a more thorough evaluation of reserves at Al-Risha field was completed. When this review showed that quantities available were not sufficient to meet the country's needs, Jordan decided to reopen talks on imports from Egypt. A pipeline was constructed and completed in 2006. Aqaba thermal power station, a major generating center, currently uses Egyptian gas. Jordan imports about 150,000 barrels of oil per day mostly from Iraq and Saudi Arabia. The Zarka refinery near Amman, the only refinery in the country, refines the imported oil. Table 2 shows electrical generation fuel consumption in 2007. Gas and oil imports pose a huge burden on the national economy. It is evident that the country is in need for renewable energy projects.

Table 1. Energy production in 2007 by generation type [6]

Fuel Type	Generating Plants Capacity (MW)	Electrical Production (GWh)
Steam Units	1013	6,904
Gas Turbines/Diesel	193	45
Gas Turbines/Natural Gas	310	916
Diesel Engines	43	1
Hydro Units	12	61
Windmills	1.4	3
Biogas	4	10
Combined Cycle	600	5,061
Total Generation	2176.4	13,001

Table 2. Fuel consumption in 2007 for electrical generation [6]

Fuel Type	Consumption in Thousands Tons Oil Equivalent
Heavy Fuel	621
Natural Gas	2,396
Diesel	9
Total	3,026

According to 2007 data supplied by NEPCO [6], electrical production cost is US$0.073 per KWh: out of which fuel cost is US$ 0.0386 per KWh. This figure is considered expensive as compared to production cost in other countries. As the oil prices surged to more than double in the summer of 2008, accordingly the production cost increased to US$0.11 per kWh. If it is assumed that the true value of oil price is $100 per barrel, the production cost would be about US$0.095 per kWh. This figure will be used in cost comparison.

ASSESSMENT OF WIND AND SOLAR ENERGIES

In Jordan electricity demand grew at the rate of 9.7% in 2007. The Jordanian government has been seeking ways to attract foreign capital to fund additional capacity. Wind and solar energies as main source of electricity generation are currently set as government priorities. The government implemented the following actions [7]:

- Developing new wind and solar maps for Jordan.
- Developing a legal framework for renewable energy.
- Developing incentives for renewable energy projects.
- Securing appropriate funding to implement the first commercial wind energy project in Jordan.
- Secure appropriate funding to implement the feasibility study of the hybrid solar power plant.

Because of the government enthusiasm to promote renewable energy, a thorough investigation has been conducted to study the possibility for a hybrid system of windmills and solar arrays for electricity generation. Data collected over many years by the JMD [8] has helped in locating the sights for both windmills and solar arrays.

The wind farm location was set in the area of RasMunif where the annual wind speed average is 5.5 meters/sec according to data collected by the Meteorological Department in Jordan. With the exception of the months of September and October, where the wind speed is low, the other 10 months the speed varies from 6 to 6.5 meter/sec. This speed represents the village ground level speed. If the windmills are sighted at higher elevation and in the valley curvature between mountains running west to east where wind tunnel effect exists, the average wind speed might rise to comfortable levels where windmills run near full capacity. The windmill tower height of 100 meters also increases wind speed to levels close to the 7-9 meter/sec [9], which might bring the wind turbine power output to 1.0 MW or more. Experience indicates that wind speed tend to be higher during the nighttime. Therefore, during the daytime the deficiency in windmills' power output can be augmented by solar cells. The average sunshine hours throughout the year are 8.5 hours per day. In the summer months, May through September, the average sunshine hours are more than 10 hours per day. In winter months,

November through February, the average sunshine hours are about 6 hours per day. This means that the solar arrays can supplement the wind turbines daily for 10 hours in the summer and 6 hours in the winter months. The solar arrays can be scattered between the windmill towers.

THE WINDMILL-SOLAR HYBRID SYSTEM

Proposed Windmill-Solar Hybrid

The proposed non-conventional electrical generation should supply 100-150 MW. As it was pointed out earlier, the sight is chosen in a high valley curvature in the mountainous range where wind tunnel effect exists when continuous high-speed wind prevails all year round. An array of 100 windmills was chosen for this work. Each unit has a capacity 1.5 MW. Several windmill suppliers were investigated and the choice was set on SAIP Electric Group [10]. Figure 1 shows the windmill chosen for this project. Since the average annual wind speed characteristics at location is 6 meter/sec and might average about 8-9 m/sec as was pointed out in Section 5.0 above.

The cut-in wind speed is 3 m/sec, which is way above the annual average wind speed guaranteeing continuous power output. The cutout speed is 25 meter/sec where this average is over 10 minutes on the average. In that location wind speeds never reach that high. Figure 2 indicates that the windmill average power output would be about 1.0 MW for year round.

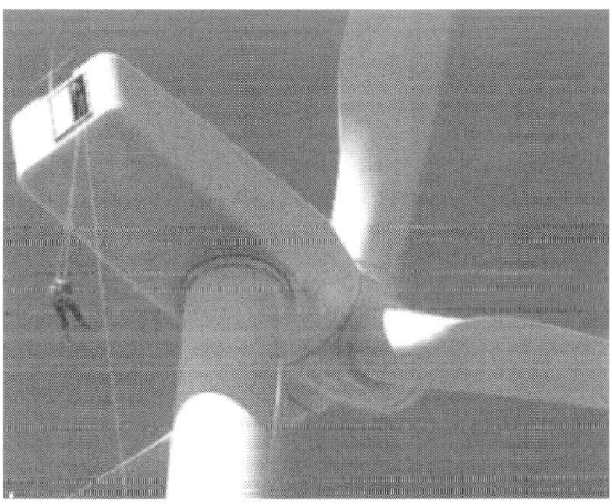

Figure 1. The 1.5 MW windmill

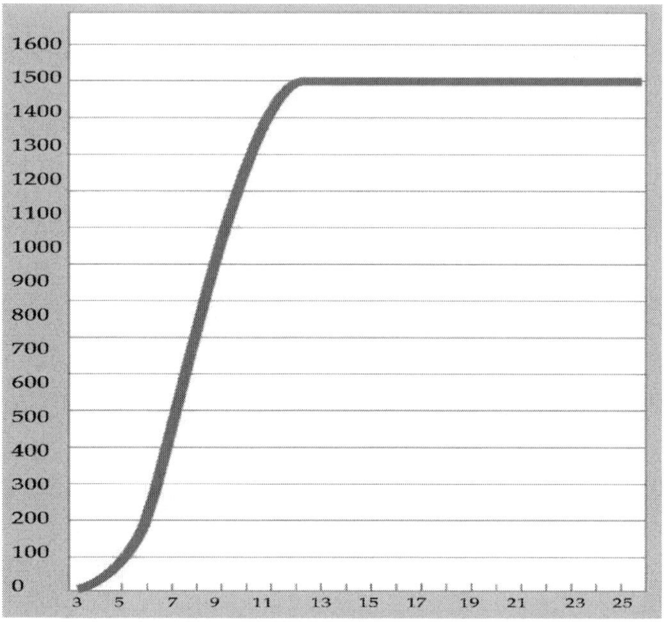

Figure 2. The power-wind speed characteristics

This power may increase up to 1.5 MW, which is the maximum power output of the generator. Therefore, the proposed windmills farm may produce a continuous power output between 100-150 MW. The blade length is 37.5 m, making the windmill side clearance 75 m, and at a height of 100 m. Leaving additional side clearance of 100 m so that windmills do not form wind obstacles between each other. Therefore, the wind farm array farm should be about 2 km long. Land appropriation for this sight would be about 200,000 m^2. In case the windmills power output is reduced, solar cells array may be an alternative for additional support and reliability. Experience had told us that whenever the wind speed drops means a fair weather where the sunshine is a maximum.

Table 3 shows the solar array type specifications to be used. The decision was to install solar array to produce 100MW to support the windmill array. A total of 500,000 arrays are needed to supply the required power of 100 MW. The array will be installed in the lower area in the flat planes.

Table 3. Solar module specifications

Maximum power (Wp)	*200W*
Maximum power voltage (V)	42
Maximum power current (A)	5.24
Open circuit voltage (V)	50
Short circuit current (A)	5.7
Number of cells (Pcs)	91(7x13)
Size of module (mm)	1702x945x45
Weight per piece (kg)	19

Since Ras-Munif, the location of the windmills is mountainous area, is also suitable for solar arrays but limited because it can also be used for agricultural plantations. A better location of the solar modules is in the Easter Jordanian Desert. In the desert, land is readily available and the yearly average daily sunshine is 9.3 hours. By installing east-west sun tracing system, a full 8 hours daily average maximum power output can be obtained. Accounting for modules surface area and spaces between modules, the solar installation requires land appropriation of 1.0 km^2. Location for the solar power station is chosen to be near Al-Risha Power Station currently in operation in the Eastern Jordanian Desert.

The sight is chosen for easier link to the national grid. Al-Risha is located at 32° latitude. This requires the modules to be installed inclined toward the South at 32° with the horizontal; facing southward Modules inclination adjustments of ±15° are needed to track the sun's seasonal variations. East-west tracking motors may be used to increase full capacity power production to 9 hours per day. The proposed hybrid wind-solar installation is needed to supply Jordan with low cost renewable electric power. These two installations are capable of supplying 10% of the country's electricity peak demand needs for the year 2009.

Cost Estimation

The cost of one windmill is US$ 1.85 million [10]; additional 20% of the price may cover shipping and installation. In addition, US$200,000.00 per unit for controllers and other supporting equipment may be needed for grid link. The total cost per unit ready to supply the grid may be set around US$2.4 million. Another 10% for maintenance, 6% for capital investment, and 5% for administration to be added to the US$2.4 million; results in net cost of US$2.9 million per unit. A total of US$290 million are needed for the wind farm installation to produce 100–150 MW of electrical power. The average lifetime of the windmill is 20 years. Simple calculations, after the assumption that the full wind power output is for 20 hours per day, result in electrical production cost of US$0.02/kWh.

As for the solar power station, the cost of 500,000 modules needed to produce 100 MW is US$370 million. This cost includes the controllers for the ground link. In addition to the US$370 million solar modules cost, 20% for installation, 10% for the sun tracking, 6% capital investment, and 5% for administration, bringing the total cost for the solar power station to US$560 million. The high percentage for installation is to cover the cost of frames upon which the modules will be installed. Remembering that the power production is for 8 hours per day and the lifetime of the solar cells is 25 years; the production cost will be US$0.077/kWh. This production cost is almost the same as the present conventional cost but lower than the projected cost of US$0.09/kWh when the price of oil reaches US$100 per barrel. Comparing the wind and solar power production costs, it seems that wind power production cost is lower than the solar power production. Therefore, wind energy production is feasible now even with oil prices at US$40 per barrel. In the future when oil prices rise, even with the high cost of solar energy, solar power is important for power floe reliability.

RAS-MUNIF – VILLAGE OF 'EBELIN HYBRID POWER PROJECT

Ras-Munif, located in the province of Ajloun, is one of the highest mountains in Jordan: about 1198 m above sea level. The villages of 'Ebelin are also located in the province of Ajloun and 4.9 km from the city center of Ajloun, directly below Ras-Munif, see Figure 3. Land terrain, wind characteristics, solar sunshine days, and politics are considered important issues for the location of wind and solar farms. It is clear that the wind speed and sunshine days data mined by JMD make Ras-Munif an ideal location for wind and solar farms.

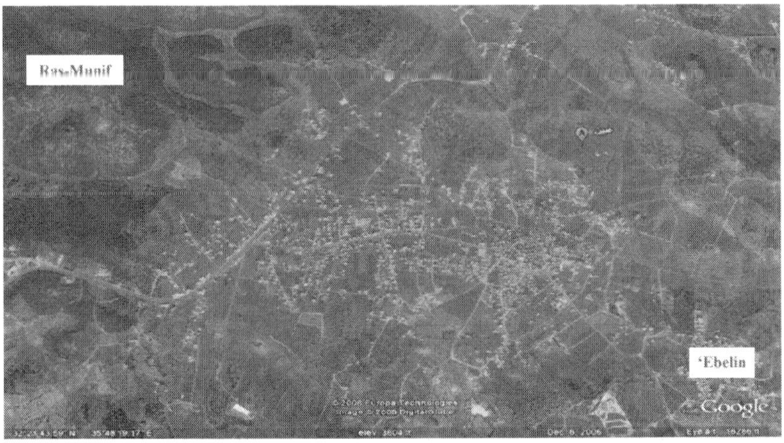

Figure 3. Area view of Ras Munif and 'Ebelin

Figure 4. Terrain location of wind turbines

Wind Energy

Wind energy depends mostly on wind speed and kinetic energy of the air mass even though wind speed is also affected by air density, air temperature, air barometric pressure, altitude, and local terrain. Wind generators are practical where the average wind speed is greater than 4.5 m/s and with constant flow rate at minimum turbulence and minimum powerful wind bursts. The Figure 4 shows best wind turbine locations on mountainous terrain.

JMD has kept data of wind speeds and directions at different locations in Jordan; wind speed and direction are very important factors for location of wind farms. Ras-Munif, one of highest mountains in Jordan, is very rich in wind speeds with an average wind speed of 5.5 m/s reaching a maximum value of 6.37 m/s during winter.

Table 4. Kyocera (kc130gt) solar panel rating

Price Per Panel	431.87JD	Short Circuit Current	8.02 A
Rated Power	130 W	Area of Panel	0.929 m^2
Rated Voltage	17.6 V	At 70% Peak Load	5385
Rated Current	7.39 A	Cells Per Panel	36 Polycrystalline Cells
Open Circuit Voltage	21.1 V	Cost Per KW	US$4681.34/KW

Solar Energy

Solar energy depends on the amount of direct sunlight even though clouds, blue patches, shades, and rain also affect direct sunlight. Solar panels are located at areas with best sun exposure. Solar panels are practical where the average sunshine is greater than 5 hours a day. The solar panels must be kept cool with minimum clouds (equivalent to approximately 50% peak sun), minimum blue patches and shades (shading even one cell of the panels can reduce the output of an entire array), and less rain (equivalent to approximately 20% peak sun). JMD has kept data on sunshine hours at different locations in Jordan. Ras-Munif is very rich in sunshine due to its elevation [7–8] with average sunshine of 8.5 hours per day and reaching a maximum of 12.3 hours/day during summer.

Hybrid Power System

In addition to the present conventional power system, the hybrid power system of Ras-Munif consists of solar panel arrays and generators; the hybrid system is tied in to the conventional system. The output voltage of the solar arrays and the wind generators are tied and synchronized together with the conventional power system main bus at the same potential. The voltage at the main bus is kept constant and used to supply the load. The Figure 5 [7–8] shows the location of the hybrid power generation system located at Ras-Munif, and transmission lines from Ras-Munif to 'Ebelin villages. The power produced at 'Ebelin is at 230 V. The solar panel selected and built into the solar arrays, is the low cost Kyocera module KC130GT with rating shown in Table 4.

The solar panels are connected in two format arrays – serial and parallel. The solar panels are connected in series to meet the voltage requirements and in parallel to meet the current requirements. The output of the arrays has a DC/DC converter linkage integrated into voltage-source-inverter (VSI) system to hold the voltage at constant value. The solar panels have sun trackers for adjusting the panel tilts during winter and summer according to the following equations [7–8]:

$$\text{Summer Tilt} = \text{Location Latitude} - 15^\circ \quad (1)$$

$$\text{Winter Tilt} = \text{Location Latitude} + 15^\circ \quad (2)$$

The Ras-Munif site selected has "Location Latitude" of 32° with the horizontal facing south. All the wind generator turbines are the horizontal axis wind turbine types because they are low cost and easy to maintain. The smallest wind generator is rated at least 10 KW. The characteristics of a typical generator are shown in Table 5.

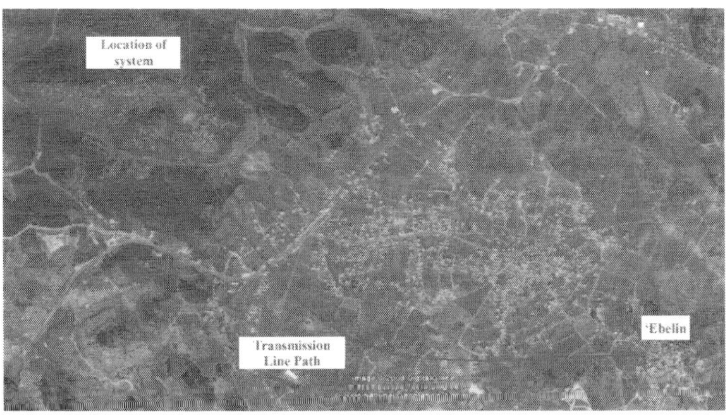

Figure 5. Location of hybrid power system and transmission line path

Table 5. Typical wind turbine characteristics

Power Rating	>10 KW	Blade Material	Fiberglass-reinforced Epoxy
Rated Wind Speed	13 m/s	Yaw Control	Passive Aligned by Tail vane
Cut-in Wind Speed	3 m/s	Rate RPM	25 - 300
Rotor Diameter	> 7 m	Tower Height	12 – 40 m
Number of Blades	3	Weight	Minimum 7540 kg

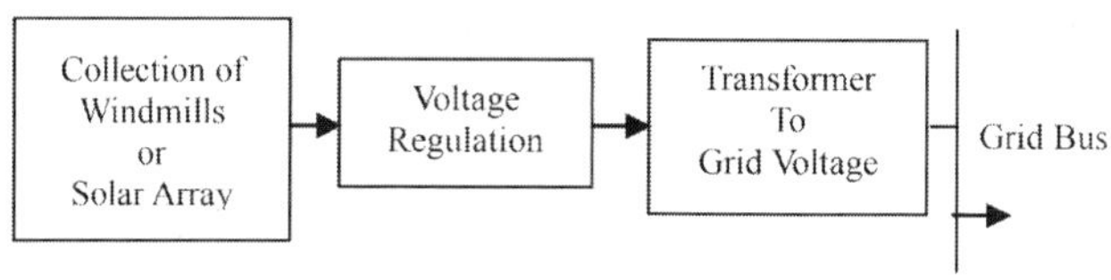

Figure 6. Proposed power system

POWER ELECTRONICS

Electrical and Power Requirements

The solar- and wind-energy systems must be in into the national grid. A block diagram of the proposed is shown in the block diagram of Figure 6. Papers [11–20] have presented a converter topologies for wind generators for wind energy conversion systems. There are various power electronic converters that have been developed. Power Electronics converters that have the maximum power and to allow for variable speed operation of wind turbines. In this project the power electronics converter selected has the following properties: 1) the maximum power obtained from control system of the converter is compared with the maximum power point tracking (MPPT) curve [18–21] at wind speeds/sun shine levels, 2) the converter must provide the required residential/commercial voltage, 3) the converter must provide frequency to within the specified error, and 4) the efficiency for small scale power must be met. A permanent magnetic synchronous generator (PMSG) and a supply-side voltage source inverter (VSI) are selected for the wind-energy for lower cost and higher power output. The DC/DC–VSI combination is capable of handling weak sun AC systems. There are various control strategies for the VSIs including d-axis and q-axis PI controllers and use of space vector modulation (SVM) to achieve a better modulation index. Even though PMSGs have initial higher cost (price of magnets) and may be demagnetized (high temperatures, overloading, and short circuits), they are flexible, have high output power without the need to increase size of generator, have lower maintenance costs (no carbon brushes, bearings, etc.), have lower losses, have very high torque at low speeds, are self-exciting machines, and do not need cooling systems. A major advantage of using PMSG is that they do not require external excitation and simple diode rectifier circuit may be used at the generator terminals. Figure 7 shows the block diagram of the solarand wind-energy system.

The solar system includes a boost converter for the MPPT. A DC/DC converter linkage is integrated into the VSI and solar systems 1) to control the generator-side DC-generator for both the solar and wind systems, 2) to maintain the desired DC-voltage for the inverter-side, 3) to eliminate certain harmonics, and 4) to provide more flexible control. The wind-generator output power is maximized using MPPT control systems and algorithms. The p-q theory [22] is used to control the active and reactive power, and the power factor.

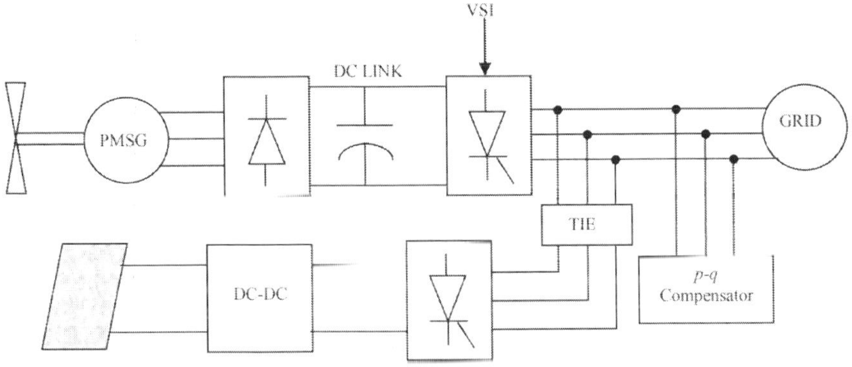

Figure 7. Hybrid solar- and wind-energy power system

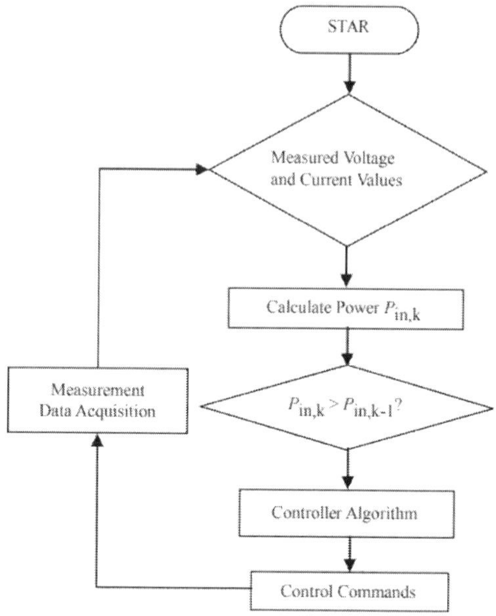

Figure 8. MPPT control algorithm process

The MPPT can be used to achieve optimal operation mode of the solar- and wind-generator power conversion system. The MPPT does not require knowledge of optimal power characteristics or measurement of wind speed, does not depend on the rotor-speed rating of the wind-generator, and does not depend on the power rating of the DC-DC converter. The required voltage and current signals are measured using sensors such as Hall Effect or linear electro-magnetic (LEM) sensors through analog-to-digital (ADC) converters. The Figure 8 shows a generic representative flowchart of an MPPT control algorithm. In the flowchart the MPPT battery system is not taken into consideration. The error signals are obtained by comparing the reference control signals and corresponding measured values. The MPPT algorithm is then applied to the error signals. The duty-cycle ratio change command is then implemented.

CONCLUSIONS

Jordan has high electric production cost that is directly linked with oil prices. An alternative is renewable wind and solar electric power production. The possibility was thoroughly investigated. The result is to install windmill farm in the mountainous area in the north, where wind speed proved to be viable, while the eastern desert is suitable to install solar power station. The cost for the windmill farm to produce 100 – 150 MW for 20 hours per day is US$290 million. The cost of the solar power station to produce 100 MW for 8 hours per day is US$560 million. The production cost is US$0.02/kWh for the windmill and US$0.077/kWh for the solar. The conventional production cost is US$0.095/kWh projected when the price of oil is US$100 per barrel. For reliable energy system, hybrid power production is essential.

The features of the generator-converter are considered to meet the requirements for the wind and solar systems. The solar- and wind-generator power outputs can be m per day is US$290 milli aximized using MPPT control systems and algorithms. The p-q theory is used to control the active and reactive power, and the power factor.

REFERENCES

1. N. Kodama, T. Matzuzaka, and N. Inomita, "power variation control of a wind turbine using probabilistic optimal control, including feed-forward control for wind speed," Wind Engineering, Vol. 24, No. 1, pp. 13–23, January 2000.
2. L. L. Freris, "Wind energy conversion systems," Englewood Cliffs, NJ, Prentice-Hall, pp. 182–184, 1990.
3. E. Koutroulis and K. Klaitzakis, "Design of a maximum power tracking system for wind-energy-conversion applications," IEEE Transaction on Industrial Electronics, Vol. 53, No. 2, pp. 486–494, April 2006.
4. E. Muljadi and C. P. Butterfield, "Pitch-controlled variable-speed wind turbine generation," IEEE Transaction on Industry Applications, Vol. 37, No. 1, pp. 240–246, January 2001.
5. W. Lin, H. Matsuo, and Y. Ishizuka, "Performance characteristics of buck-boost type two-input DC-DC converter with an active voltage clamp," IEICE Technique Report, Vol. 102, No. 567, pp. 7–13, January 2003.
6. http://www.nepco.com.jo/showImageTC.aspx?imageURL=Statistics_files/Englishalbums/2/, Retrieved on 2/3/2009.
7. http://www.ren21.net/iap/commitment2.asp?id=93, Retrieved 3/3/2009.
8. The Hashimite Kingdom of Jordan, Meteorological Department, Climate Division, Jordan Climatic Data, 2007.
9. E. W. Peterson and J. P. Hennessey Jr., "On the use of power laws for estimates of wind power potential, Journal of Applied Meteorology, Vol. 17, 1978.
10. SAIP Electric Group Limited, Huifeng Road, Luishi Industrial Zone, Wenzhou, Zhejiang, 325604, China.
11. J. A. Baroudi, V. D. Dinavahi, and A. M. Knight, "A review of power converter topologies for wind generators," Renewable Energy 32, Science Direct, pp. 229–238, January, 2007.
12. Z. Chen and E. Spooner, "Current source thyristor inverter and its active compensation system," Proceedings of IEE Generation, Transmission, and Distribution, Vol. 150, pp. 447–454, July 2003.

13. K. Tan and S. Islam, "Optimum control strategies in energy conversion of PSMG wind turbine system without mechanical sensors," IEEE Transaction on Energy Conversion, Vol. 10, pp. 392–399, 2004.
14. Z. Chen and E. Spooner, "Grid power quality with variable speed wind turbines," IEEE Transaction on Energy Conversion, Vol. 16, 2001, pp. 148–154.
15. Z. Chen and E. Spooner, "Wind turbine power converters: A comparative study," Proceedings of IEE Seventh International Conference on Power Electronics and Variable Speed Drives, pp. 471–476, September 1998.
16. S. H. Song, S. Kang, and N. Hahm, "Implementation and control of grid connected AC-DC-AC power converter for variable speed wind energy conversion system," Proceedings of IEEE AIPEC'03, Vol. 1, pp. 154–158, February 2003.
17. Y. Higuchi, N. Yamamura, M. Ishida, and T. Hori, "An improvement of performance of small-scaled wind power generating with permanent magnetic type synchronous generator," Proceedings of IEEE IECON'00, Vol. 2, pp. 1037–1043, October 2000.

4

An Overview of Research on Optimization of Integrated Solar/Wind Power Generation Systems

Zhonglei Shao, Kwok Lun Lo

Department of Electronic and Electrical Engineering, University of Strathclyde, Glasgow, UK.

ABSTRACT

Although transmission systems are able to cover most of the areas in many countries, there are still some isolated areas such as rural counties and remote desert lands where grid power cannot be accessed. Therefore, a reliable and economical power supply scheme is required to solve the problem. One of them combines wind/solar power generation with the support of storage system. This paper is to give an overview of the optimization methodologies about the wind/solar stand-alone system supported by storage systems or integrating with other renewable or conventional power generation sources. It is shown that continued research and optimization methodology in this area are still in great need for performance improvement.

Keywords: Solar/Wind Generation System, Optimization, Techno Eco nomic

Citation: Shao, Z. and Lo, K. (2016), "An Overview of Research on Optimization of Integrated Solar/Wind Power Generation Systems". World Journal of Engineering and Technology, 4, 35-42. doi: 10.4236/wjet.2016.43D005.

INTRODUCTION

As the global environmental pollution becomes more and more serious, distributed power generation system based on renewable energy, such as photovoltaic and wind power generation, attracts more public attention. On one hand, due to the unstable and intermittent characteristic nature of wind and solar energy, a large capacity of energy storage is needed when the hybrid wind/solar power system (HWSPS) works in stand-alone model. On the other hand, a strong impact on the utility grid will occur when HWSPS works in grid-connected mode.

For consumers in some rural areas it may be difficult for them to access the conventional transmission system and the reason could be technological or economic issues, HWSPS would be a good alternative because of the complementary characteristics of wind and solar energy [1]. This paper will make a general review on researches that have been done on the HWSPS and also suggestion for future work.

STRUCTURE OF THE HWSPS

Figure 1 is a general block diagram about this HWSPS. Wind turbines and Photovoltaic (PV) modules function as power suppliers. AC load or DC load are included in this system connected with inverters and rectifiers. Controller collection system keeps up good performance of whole system operation by collecting data from the other parts. Data are collected from the controller part for observation to improve the performance. Either battery banks or diesel generators are added into the system in the event of insufficient power generation from the wind and solar. Excess power generated can be stored in the battery in most conditions until the battery reaches to its maximum level. Then the overproduced power must be transmitted to the dump load to be consumed for safety.

OUTPUT POWER FROM THE WIND TURBINE

Wind energy has been in use for a long time since it is clean and renewable. As a result, a way of recording wind speed accurately is highly relevant with the output power of wind turbine. It is not uncommon that the hourly records of wind speed are not available for a particular site. There are two general solutions to this problem. The first solution is to generate hourly wind speed synthetically from the monthlyaverage values of the wind speed data [1].

Another solution is to generate the wind speed data from a nearby site by making necessary adjustments [2].

The hour-by-hour simulation programs have been the main tools used to judge the long-term performance of wind energy systems. Borowy and Salameh gave a fundamental wind turbine model in [3] [4], which was further studied by other researchers [5]. The details are shown below.

Step 1: collect the wind speed data at a particular height. Most of the wind speed data are usually recorded near the ground level. These data can then be updated to the particular hub height according to formulation mentioned in [4].

Step 2: choose a proper wind speed distribution. Weibull distribution is proved to fit well with wind speed distribution. The wind speed distribution functions are calculated for each hour of a typical day in every month over a few years.

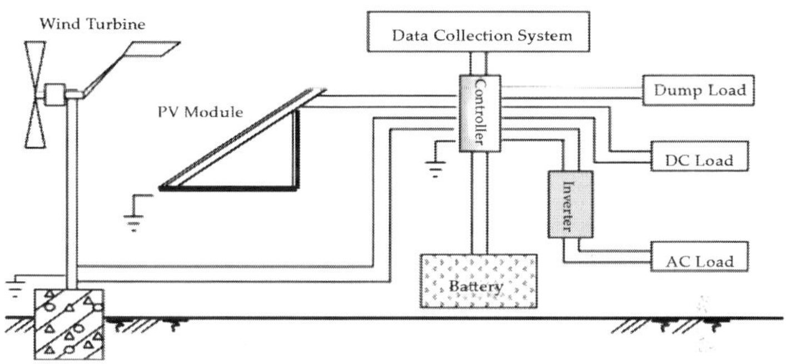

Figure 1. Block diagram of a hybrid solar? Wind power generation system [5].

Step 3: calculate the instantaneous output power from the wind turbine. The instantaneous output power of the wind turbine is a function of the wind velocity, shown in Figure 2.

Step 4: calculate the average power output of the wind turbine. Combined with the wind turbines specification provided by the manufacturer and the distribution density calculated in step 2, the average power output from the wind turbine for every hour of a typical day can be easily calculated by the following expression:

$$P_{w.avg} = \int_{0}^{\infty} P_{w}.f(v).\,dv \quad (1)$$

f(v)—the probability density function.

P_w—the instantaneous electrical power output of a specific wind turbine.

However, sometimes the hour-by-hour wind speed data are not available for some particular sites. As a result, alternative simulation programs were developed in [6] [7] to determine the long-term performance of renewable energy systems. Themore simple the simulation model is, the less accurate it would be.

OUTPUT POWER FROM THE PV MODULE

Modeling of a photo voltaic system is similar tothe wind energy system [3] [4]. Details are shown below.

Step 1: Choosea PV module tilt angle. PV modules in different locations would be adjustedto different tilt angles for the purpose of the maximum use of solar irradiance. And there are some other PV modules that can track the height of the sun and adjust the tilt angle automatically.

Step 2: Collect hourly solar irradiation data and ambient temperature. Ambient temperature data are necessary because the I-V characteristic of PV module differs with the change in either temperatures or irradiations, shown in Figure 3. When the in solation level stays the same, a PV module output power would be different due to different temperature levels. A maximum power point tracker (MPPT) is used to help the PV array to output the maximum power at any radiation level [8] [9].

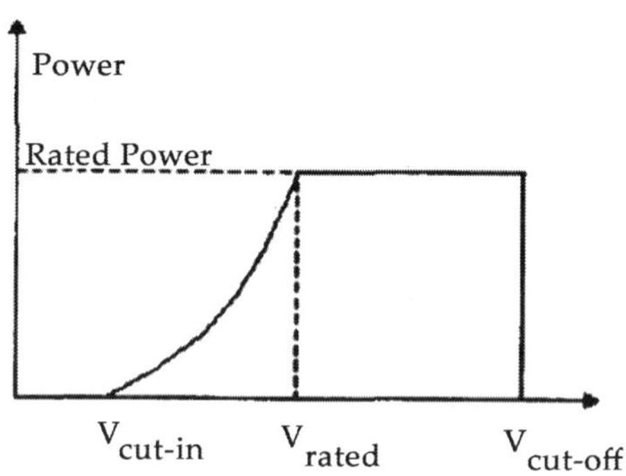

Figure 2. Power-wind speed characteristic.

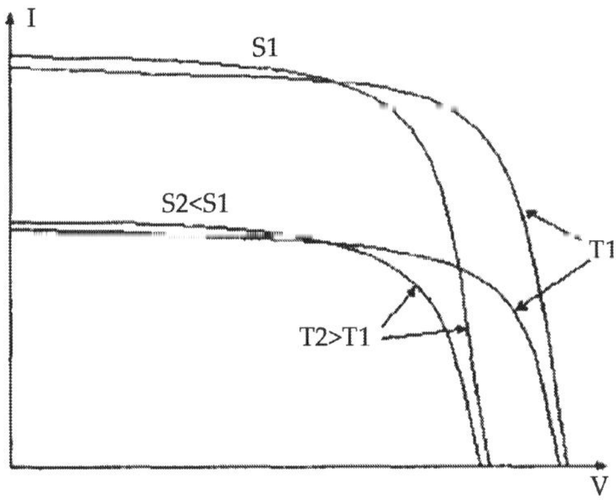

Figure 3. I-V Characteristics of a solar module for different insolation levels S1 and S2 with two different ambient temperatures T1 and T2 [4].

Step 3: Choose a proper solar irradiance distribution. Because of some insolation reducing phenomena such as different cloud cover, the solar irradiance is also a random variable which can be represented by a particular density function. Like Weibull density function is often used to estimate wind speed, Beta density function is usually chosen to describe the solar irradiance distribution [4] [5].

Step 4: Calculate the average power output from the PV module. Combined with the PV array specification provided by the manufacturerand the distribution density calculated in step 3, the average power output from the PV module for every hour of a typical day can be easily calculated by following expression [3]:

$$P_{pv,avg} = \int P(S).f(S).dS \qquad (2)$$

f(S)—irradiance probability density function

P(S)—the electrical power output of a particular PV array

Another novel model based on the I-V curves of a PV module was proposed in [10] to predict the PV module performance for engineering applications with limited data provided by the PV module manufacturers. Five parameters are introduced to account for the complex dependence of the PV module performance upon the solar-irradiance intensity and PV module temperature. It is claimed that this model is exceptionally simple and useful to calculate the actual performances of the PV modules under operating conditions.

MODELING OF BATTERY STORAGE SYSTEM

Battery storage system is necessary for HWSPS. On one hand, it contributes a lot to the reliability of the whole system. On the other hand, it reduces the amount of dump load when more power than demand is produced.

Two common conditions are accepted by most of the researchers used to analyze the operation performance of the battery. One situation is that when power generated by both the PV arrays and the wind turbines exceeds the load, the battery will enter a charging condition. The other is that when the power generated by both the PV arrays and the wind turbines is insufficient to supply the load, the battery will enter a discharging condition.

State of charge (SOC) is an important parameter that tells the current operating condition of the battery. To help to prolong battery life spans, a limited range of depth of discharge (DOD) is set which means a limited maximum power can be withdrawn from the battery. According to the specifications from the manufacturers, the battery's lifetime can be prolonged to its maximum if DOD takes the value of 30% - 50% [11]. And it is obvious that battery cannot hold any more amount of power when its capacity is reached. Then the surplus power must be transmitted to the dump load [5].

It was mentioned in [11] that only when the output voltage of PV array was higher than the float charge voltage of battery bank could the electric energy generated by PV array be used. Ref [11] introduced a way of judging whether the output of PV array was available by using a formula relating the terminal voltage of battery bank with charge quantity and ambient temperature. In practice, the way of connection of PV modules should meet the demand requirement of users both on voltage and power. The operating voltage is determined by the number of the PV array connected in series, whereas in parallel the number of battery bank strings determines the capacity of the PV array.

Umarin Sangpanich gave assumption in [12] that the less amount of energy discharged each time from a battery, the less damaging the battery storage systems would be. This process can lead to more comprehensive battery discharging process considering discharge duration by setting a range of the permitted discharging power. As a result, a different optimization algorithm of the battery management assumption considering the dynamic nature of the energy storage is posed which can lead to more future work on the battery storage system.

DIESEL GENERATOR SUPPORT

Diesel generators sometimes are also added into the HWSPS as support to improve the system reliability. It is largely due to diesel generators can provide a more stable power supply when insufficient power are produced from the wind turbine and PV array modules and the storage is depleted. One big disadvantage of diesel generators is its relatively high CO2 emission which would lead to damage to the environment.

A methodology to size an optimal stand-alone hybrid PV/wind/diesel/ battery bank using a Multi-Objective Genetic Algorithm was developed in [13]. The diesel generator model is designed for operating between 30% and 100% of their nominal power to keep a healthy lifetime. Three different kinds of diesel generators with varied nominal output power and costwere applied in [13]. With data of solar radiation, temperature and wind speed collected from Gandon, the north-western of Senegal, obtained results are presented as optimal Pareto front. Each solution from the methodology included three indices i.e. the optimal number of devices, the levelized cost of energy (LCE) and the amount of CO_2 emission. he results showed that the amount of the CO_2 emission decreased as the value of LCE increased. On one hand, local government would put money on emissions treatment. On the other hand, increased number of diesel generators means fewer numbers of PV modules, wind turbines and especially the batteries which are usually the most expensive parts, leading to less LCE.It is concluded that the use of the diesel generatorscan make an influence on the optimization of configuration. If a proper type of diesel generator is chosen then the total cost of the system would be minimized.

RELIABILITY ISSUES IN STAND-ALONE AND GRID-CONNECTED MODELS

Stand-alone HWSPS is an isolated system which only supports devices inside the system. The definition of Loss of Power Supply Probability (LPSP) is popular among researchers and has become a standard parameter used to represent the reliability of the aimed stand-alone system. LPSP is defined as the probability that an insufficient power supply results when the combined hybrid system (PV array, wind turbine, and battery) is unable to satisfy the load demand [15]. In other words, a zero value of LPSP means the load will always be satisfied and a unity value of LPSP means that the load will never be satisfied [4].

An improved optimal sizing method considering the system in both stand-alone and grid-connected operation conditions was proposed in [14]. Some objectives such as high power supply reliability, full utilization of the complementary characteristics of the wind and solar, small fluctuation of power injected into the grid and minimization of the total cost of the system are set and realized. It is shown that when the HWSPS is in a grid-connected mode, the power injected into the grid would have an impact on power quality of the utility grid. An energy filter is further applied to help smooth the fluctuation of injected power. The standard deviation (STD) and power fluctuation rate are two important parameters when considering the fluctuation of injected power. The STD is the result related with the instantaneous and average power injected into the grid during certainperiod. The power fluctuation rate is related to the maximum and minimum power injected into the grid during certain time interval. It is obvious that the smaller STD and power fluctuation rate, the smaller is the power fluctuation.

ECONOMIC ISSUES

No investors will put money into a system that is not cost-effective even when a zero LPSP condition can be achieved. As a result, from the point of an investor, not only the achieved optimum configuration should meet the power reliability requirement but also the whole system is worthwhile to be invested in [16]. The concept of life cycle cost of the system has been accepted by most researchers. This concept consists of the initial capital cost, the operation and maintenance cost and the components replacement cost. The initial capital cost is mainly from prices of PV arrays, wind turbines and batteries. The present worth value can be calculated by using the capital recovery factor which is related to the actual interest rate. Operation and maintenance cost are usually counted as a percentage part of the initial capital cost for simplicity. The replacement cost is usually included when the project lifetime (usually 15 to 20 years) exceeds the lifetime of the distributed generators [17].

When diesel generators are included in the system, CO_2 emission effect should also be accounted. It is obvious that the government's subsidy or credit also plays an important part when analyzing the economy of a renewable energy system which would lead the results to be more accurate and reasonable. What is more, a loss of power supply or a strongly fluctuant power should also be accounted when it is a grid-connected HSWPS model. It is due to the reason that fluctuant power would increase the regulation cost of the utility grid. As a result, a penalty cost is considered when the power supply reliability or the fluctuation of power injected into the grid cannot meet the requirements [11].

CONCLUSIONS

This paper has presented a general review to achieve an optimized wind/ solar hybrid system. Different models have been presented in some details including wind part, solar part and battery storage, diesel support. Important parameters are the wind turbine height and PV module tilt angle.

It is noted that not only the reliability level but also investment cost should be counted. Both stand-alone and grid-connected models have been mentioned. When it is a grid-connected condition, the fluctuation should be considered because it can increase the investment cost and has an effect on system reliability.

Further work is still needed especially in finding a new optimization algorithm for the battery management considering the dynamic nature of the energy storage.

REFERENCES

1. Knight, K.M., Klein, S.A. and Duffie, J.A. (1991) A Methodology for the Synthesis of Hourly Weather Data. Solar Energy. http://dx.doi.org/10.1016/0038-092X(91)90023-P
2. Wahab, M.A. and Essa, K.S.M. (1998) Extrapolation of Solar Irradiation Measure-ments: Case Study over Egypt. Renew Energy.
3. Borowy, B.S. and Salameh, Z.M. (1994) Optimum Photovoltaic Array Size for a Hybrid Wind/PV System. IEEE Trans. on Energy Conversion, 9, 482-488. http://dx.doi.org/10.1109/60.326466
4. Borowy, B.S. and Salameh, Z.M. (1996) Methodology for Optimally Sizing the Combination of a Battery Bank and PV Array in a Wind/PV Hybrid System. IEEE Trans. On Energy Conversion, 11, 367-375. http://dx.doi.org/10.1109/60.507648
5. Zhou, W., Lou, C.Z., Li, Z.S., Lu, L. and Yang, H.X. (2010) Current Status of Research on Opti-mum Sizing of Stand-Alone Hybrid Solar-Wind Power Generation Systems. Applied Energy, 87, 380-389. http://dx.doi.org/10.1016/j.apenergy.2009.08.012
6. Siegel, M.D., Klein, S.A. and Beckman, W.A. (1981) A Simplified Method for Esti-mating the Yearly-Mean Performance of PV Systems. Solar Energy. http://dx.doi.org/10.1016/0038-092X(81)90220-6
7. Clark, D.R., Klein, S.A. and Beckman, W.A. (1984) A Method for Estimating the Performance of PV Systems. Solar Energy. http://dx.doi.org/10.1016/0038-092X(84)90010-0
8. Kratochvil, J.A., Boyson, W.E. and King, D.L. (2004) Photovoltaic Array Performance Mo- del; Technical Report; Sandia National Laboratories: New Mexico. http://dx.doi.org/10.2172/919131
9. Soto, W.D., Klein, S. and Beckman, W. (2006) Improvement and Validation of a Model for Photo-voltaic Array Performance. Sol. Energy, 80, 78-88. http://dx.doi.org/10.1016/j.solener.2005.06.010
10. Zhou, W., Yang, H.X. and Fang, Z.H. (2007) A Novel Model for Photovoltaic Array Performance Prediction. Applied Energy, 84, 1187-1198. http://dx.doi.org/10.1016/j.apenergy.2007.04.006
11. Ai, B., Yang, H., Shen, H. and Liao, X. (2003) Computer-Aided Design of PV/Wind Hybrid System. Renewable Energy, 28, 1491-1512. http://dx.doi.org/10.1016/S0960-1481(03)00011-9
12. Umarin,S. (2012) Optimization of Wind-Solar Energy Systems Using Low Wind Speed Turbines to Improve Rural Electrification, A Thesis

for the Degree of Doctor of Philosophy at the University of Strathclyde.

13. Bilal, B.O., Sambou, V., Kébé, C.M.F., Ndiaye, P.A. and Ndongo, M. (2012) Methodology to Size an Optimal Stand-Alone PV/Wind/Diesel/Battery System Minimizing the Levelized Cost of Energy and the CO2 Emissions. Energy Procedia, 14, 1636-1647. http://dx.doi.org/10.1016/j.egypro.2011.12.1145
14. Xu, L., Ruan, X.B., Mao, C.X., Zhang, B.H. and Luo, Y. (2013) An Improved Optimal Sizing Method for Wind-Solar-Battery Hybrid Power System. IEEE Trans. Sustain. Energy, 4, 774- 785.
15. Kaabeche, A., Belhamel, M. and Ibtiouen, R. (2010) Optimal Sizing Method for Stand- Alone Hybrid PV-Wind Power Generation System.
16. Prasad, A. and Natarajan, E. (2006) Optimization of Integrated Photovoltaic-Wind Power Generation Systems with Battery Storage. Energy, 31, 1943-1954. http://dx.doi.org/10.1016/j.energy.2005.10.032
17. Koutroulis, E., Kolokotsa, D., Potirakis, A. and Kalaitzakis, K. (2006) Methodology for Optimal Sizing of Stand-Alone Photovoltaic/Wind-Generator Systems Using Genetic Algorithms. Solar Energy, 80, 1072-1088. http://dx.doi.org/10.1016/j.solener.2005.11.002

SECTION 2:
SMART GRIDS

5

Energy Efficiency and Renewable Energy Technologies Using Smart Grids: Study Case on NIPE Building at UNICAMP Campus

M. D. Berni, P. C. Manduca, S. V. Bajay, J. T. V. Pereira, J. T. Fantinelli

Interdisciplinary Center on Energy Planning (NIPE), State University of Campinas (UNICAMP), Campinas, Brazil

ABSTRACT

In its broadest interpretation, the smart grid vision sees the future of power industry transformed by the introduction of intelligent two-way communications, ubiquitous metering and measurement. This enables much finer control of energy flows and the integration and efficient use of renewable forms of energy, energy efficiency methodologies and technologies, as well as many other advanced technologies, techniques and processes that wouldn't have been practicable until present. The smart grid

---

Citation: Berni, M., Manduca, P., Bajay, S., Pereira, J. and Fantinelli, J. (2014), "Energy Efficiency and Renewable Energy Technologies Using Smart Grids: Study Case on NIPE Building at UNICAMP Campus". Smart Grid and Renewable Energy, 5, 193-197. doi: 10.4236/sgre.2014.58018.

vision also enables the creation of more reliable, more robust and more secure power supply infrastructure, and helps optimize the enormous investments required to build and operate the physical infrastructure required. The smart grid promises to revolutionize the electric power business that has been in place for the past 75 years. This work discusses the efficiency, targeted at the consumer units of electricity, with a view to sustainability and potential for technological innovation. The issue is addressed from two perspectives: the systems for generation and power distribution, and the design of a building "smart energy". Because of the novelty of the subject in our country, the concepts presented and treated throughout this work come from material obtained at events and specialized sites on electric power system in Brazil and worldwide, being accompanied by information and data from NIPE's building at University of Campinas's campus case study in which it exemplifies the applicability of the techniques and recommended technologies.

Keywords: Smart Grids, Energy Efficiency, Renewable Energy, Smart Building, Generation and Distribution System, Decentralized Generation

INTRODUCTION

The ongoing structural and technological changes make it possible to expand the renewable energy use, decentralized generation and energy efficiency in order to supply the demand for electricity. In this way, besides the fundamental concepts of operation, transmission and distribution knowledge on electrical systems, it requires to include topics such as information and communications technologies, and signal processing to improve energy used by smart grid (SG).

SG is becoming a reality in many developed countries like United States, Japan and Germany with the implementation of pilot projects countries and improvement actions in their supplying networks [1] . Brazil's Federal Government establishes the guidelines of the Brazilian Program of Smart Grids (PBREI) through the National Electric Energy Agency (ANEEL) [2] and the 482/2012 Resolution paves the way to replace 67 million conventional electricity meters for smart meters. Universities, corporations, NGOs, government and society worldwide had been concerned with energy issues. The energy field involves many aspects of social life including the background on engineering, environment, logistics, sources, and, more recently the use of information technology (IT) "intelligence" to optimize and manage power systems, adjusting supply, demand and efficiency in the use of electricity.

Modern life has made electricity an increasingly vital product. In any segment such as production of goods or services for the public safety, health or simply for the comfort of homes, electricity is an indispensable element. However, its intense use stresses the system and request for more power production. Whereas the world is generating hydro, an equally increasing impact on the environment has been seen. Environmental issues have reached their limit and have the most importance for the mankind survival and that is the reason why to think of solutions that reconcile energy production and environmental preservation. In this context, the point of view of the power distribution system and the future of the consumer unit, covered by SG concept, will be prioritized. Utilities and consumers will converge their interests and make use of new technologies to achieve energy efficiency, reliability on electricity distribution systems, decrease cost and reducing environment and natural resources. This work discusses the efficiency, targeted at the consumer units of electricity, with a view to sustainability and potential for technological innovation. The issue is addressed from two perspectives: the systems for generation and power distribution, and the design of a building "smart energy". Furthermore, this work discusses a case study on NIPE Building at UNICAMP campus, which has good values to the scientific community and contributed new information to the related field. Finally, Table 1 provides an example project approach for analysis energy efficiency and smart grid for this case study.

RENEWABLE ENERGY AND ENERGY EFFICIENCY ON BRAZILIAN CONTEXT

In Brazil energy efficiency is clearly less important than the addition of "new energy" to the grid, despite the great potential of reducing energy intensity of the Brazilian GDP and the recent successful experience in increase energy conservation during the blackout in 2000. Making energy efficiency the key topic in the whole society should be priority in the government agenda. The little importance given to the subject in the Ten Year Energy Plan in 2019 [3] , neglects the fact that investments in the area are smaller and faster return and therefore should be highlighted in the policy and government plans, especially when the country should observe growth of energy demand of 54 GW over the next ten years.

Renewable energy together with energy efficiency alternatives should be considered because the wide social and environmental benefits that often result to generate electricity. Brazil has several options for generating clean and competitive energy for its expansion: hydropower, biomass, wind and solar power. In less developed countries, there are also ocean energy and geothermal energy. But as important as increasing the supply of renewable energy is to increase the efficiency of energy consumption generated from renewable sources or not in the economic fields. Moreover, the improving of energy efficiency can occur in more narrow term to transition to a more renewable energy sources, which will not happen abruptly, since the dynamics that sustain the current non-renewable energy model cant not be easily reversed for many reasons such as: (i) the high level of material consumption and energy in an emerging country, (ii) the non-renewable energy infrastructure already established, (iii) the growing demand for electricity services by applicants, and (iv) population growth.

SMART GRID AND ENERGY EFFICIENCY

SG represents the application of IT in the integrated communication electrical system. This technology involves installing sensors on the lines of the electric power grid, embedded with chips that detect data on the operation and performance of the network like voltage and current. The devices analyse those information to determine what is significant. For example, if the voltage is too high or too low. When the sensors detect significant information communication data to a central analytical system where it will analyse them and determine what is wrong and what should be done to improve network performance occurs. In the case of very high voltage, the software will instruct one of the devices already installed in the network to reduce the voltage, thus saving energy.

Table 1. Phases in a sustainable building retrofit.

Phase 1	Phase 4
Project setup	I & C
definescopeofwork	implementation
availableresources	comissioning
prediagnostic	Phase 5
Phase 2	Validation

Energy auditing	verification M & V
selectindicators	
measures	
finaldiagnostic	SUSTAINABLE
Phase 3	BUILDING
Retrofitoptions	
energysaving	energyefficiency
economicanalysis	and
energygenerationon site	smart grid

SG technology has been benefits with fuel efficiency, which means less energy the utility company to provide equal or better quality of service to its customers; increases the reliability of the electric system, reducing costs and carbon emissions. The SG detects when the assets of a network fail or are with declining performance, will identify them to the concessionary can repair them or replace them before there is a real power outage, but also allows you to isolate the impact of a failure to customers, so that fewer customers are affected when there is a power failure. Lastly is the integration of cutting edge , ranging from reading a smart meter system to interact with the client's management at home, solar panels , which require interaction with the network to achieve success [4] . Furthermore, the SG lever distributed generation from renewable sources, in that power generation can be carried out at or near the independent power consumer, technology and energy source. The main advantage of distributed generation systems is the savings in investments, transmission and reduction in losses in these systems, enhancing the stability of electric power service and increases energy efficiency.

With the SG tool it can relax the retrofits. In the energy context, retrofit is used to define changes and upgrades in systems generators and consumers of electricity aimed at their conservation. This type of application occurs in power plants with reform or adding equipment to increase efficiency, production and life. A common case is the retrofit of boilers in power plants. Another important example of retrofit is the object of the case study presented below in the NIPE's building at University of Campinas's campus. Retrofit buildings is normally associated with the change in lighting systems, electrical and plumbing installations through energy efficient communication technologies and advanced quality. In addition to improving energy efficiency, retrofit helps to reduce emissions of greenhouse gases over the life cycle of buildings.

RETROFIT BUILDINGS: BRIEF DESCRIPTION OF THE CASE STUDY PROGRESS IN BUILDING NIPE-UNICAMP

Campus at UNICAMP as same micro representatives of society contains buildings with a variety of functions and uses, each with their own characteristics in the consumption of energy, water and waste. As a task to become "good examples" for society in general, universities need to develop "best practices" to make sustainable buildings, both in the rehabilitation of buildings and in new buildings. Sustainable buildings on campus may have educational goals for college students and the academic community, reflected in potential "best practices" for society at large gains.

Energy efficiency and renewable energy in the built environment stand out as one of the main alternatives to minimize problems arising from climate change. The retrofit project of NIPE-UNICAMP will demonstrate concrete gains on rehabilitation of a building live "post-occupation" by proposing actions to saving energy, maximizing the educational impact of these actions and awareness through interaction with students and the university's community, under the supervision of experts. Participatory nature of the project will allow the perception of specific possibilities of a green building among users of the Campus, increasing local knowledge to be adapted to the "perfecting" of the sustainability process extra-campus. The project will share know-how to process a wider spread in the society. Against this background, the NIPE-UNICAMP project deals with the theme "Smart Grid and Sustainable Buildings" with the main objective of the study and evaluation of opportunities for energy efficiency and comfort environmental post-occupancy building for research, considering the surroundings, air conditioning system, lighting type and acoustic quality, based on the evaluation of sustainability of the built environment through possibilities INMETRO nota certification. Therefore, it is a survey of constructive characteristics, architectural and the thermal NIPE-UNICAMP, followed by the determination of energy needs through software simulation, and finally evaluating the results and proposing alternatives for energy optimization and environmental comfort NIPE through the retrofit of the building interventions from the sustainability approach, involving the energy performance, thermal quality , lighting quality and acoustics quality, should integrate the agendas of pilot projects and retrofits that begin to be stimulated and developed by various national and international organizations that study the impacts of the sector building and construction for society and the environment [5] .

There are many softwares available for the study and evaluation of opportunities for energy efficiency and environmental comfort, and mention: i) DOE2 (EQUEST), ii) Design Builder, iii) ECOTEC, iv) TRNSYS, v) and vi) Energy Plus and ESP-r, among others [6] . Besides these, there RET Screen software support for decision making opportunities for energy efficiency with a focus on renewable energy to a building. This is free software, provided by the Government of Canada, which helps the decision maker to identify and access potential energy projects that bring reduction in energy intensity of a building, including its technical and economical viability [7] . Regarding the thermal comfort of the environment, their qualification and quantification requires making measurements, for example if the air temperature and the Globe. From measurements of the thermal evaluation can be performed by the Predicted Mean Vote (PMV), while simulations with 2:03 Comfort software method. Regarding the study and analysis of the luminal comfort, it has been alternatively windsurfing and Reluxcad and finally, evaluation of acoustic comfort in the Brazilian case can be according to NBR 10152 [8] .

FINAL CONSIDERATIONS

There is a strong correlation between energy consumption, environmental comfort and the life cycle of a building. Indicators of energy intensity throughout the life cycle analysis of a building provide a way to understand the evolution of that correlation. The energy intensity can be reduced in two ways. First, greater energy efficiency can reduce the energy consumed to produce the same level of energy services (for example, a more efficient light bulb produces the same light with less energy consumption). Second, the issues surrounding sustainability, markets and peer pressure, end up imposing changes in energy intensive activities, such as the search for the lowest energy consumption activities, greater comfort and environmental activities and/or less carbon intensive process. Energy efficiency assisted by replacing fossil energy by renewable and sustainability criteria in the lifecycle is the key to driving incremental reduction in energy intensity and can offer solutions as diverse as climate change, energy security, competitiveness, and human being and socio-economic development.

M. D. Berni, P. C. Manduca, S. V. Bajay, J. T. V. Pereira, J. T. Fantinelli (GHG) is 30% [9]. The final energy consumption in Brazil, only for residential household, more public trade, represents 47.1% of the total consumption of energy sources [10] . The evaluation of energy efficiency and environmental comfort in a built environment, post-occupancy, as NIPE-UNICAMP, is requesting a careful study to obtain data that allow integrate simulations and diagnostics to make decisions.

REFERENCES

1. IEEE PES (2013) http://smartgrid.ieee.org/resources
2. Silveira, P.M. and Ribeiro, P.F. (2012) Introdução do Conceito de Redes Elétricas Inteligentes no Currículo do Engenheiro Eletricista Brasileiro. XL Congresso Brasileiro de Educação em Engenharia, Belém, Pará, Brazil.
3. EPE (2013) Empresa de Pesquisa Energética, Plano Decenal de Energia 2019, Ministério de Minas e Energia. http://www.epe.gov.br/imprensa/PressReleases/20100504_2.pdf
4. GVces, Centro de Estudos em Sustentabilidade, Fundação Getulio Vargas (FGV) (2011) Fontes de Energia e Eficiência Energética, Plataforma Empresas pelo Clima, Oficina de Trabalho IV, 17 p.
5. UNEP-SBCI (2012) United Nations Environment Programme—Common Carbon Metric: Protocol for Measuring Energy Use and Reporting Greenhouse Gas Emissions fro Building Operations. www.unepsbci.org
6. EERE (2013) Energy Efficiency and Renewable Energy, Building Energy Software Tools Directory. http://apps1.eere.energy.gov/buildings/tools_directory/alpha_list.cfm
7. (2012) http://www.retscreen.net
8. Kowaltowski, D.C.C.K., Fávero, E., Borges Filho, F., Gouveia, A.P., Ruschel, R.C., Pina, S.A.G. and Gomez, V.S. (2001) Ensino do Projeto Arquitetônico: A Teoria Traduzida em Exercícios no Ensino do Projeto Arquitetônico. Revista da Escola de Minas, Ouro Preto, Vol. 54, 1-6.
9. UNEP-SBCI (2013) United Nations Environment Programme—Sustainable Building & Climate Iniciative. www.unepsbci.org
10. EPE BEN (2012) Balanço Energético Nacional ano Base 2011, Empresa de Pesquisa Energética (EPE), Ministério de Minas e Energia (MME), Brasília, DF.

6

A Perspective on the Future of Distribution: Smart Grids, State of the Art, Benefits and Research Plans

Rosario Miceli, Salvatore Favuzza, Fabio Genduso

Dipartimento di Ingegneria Elettrica, Elettronica delle Telecomunicazioni, di Tecnologie Chimiche, Automatica e Modelli Matematici, Università di Palermo, Palermo, Italy

ABSTRACT

Currently, the design and operation criteria for electrical distribution networks are fastly changing due to some factors; among these, the progressive penetration of Distributed Generation (DG) is destined to cause deep changes in the existing networks, no longer considered as passive terminations of the whole electrical system. Moreover, the increasing application of Information Communication Technologies (ICT) will allow the implementation of the so called "smart grids", determining new interesting scenarios. In the paper the problems and the potential benefits of DG, the possible new electrical distribution system models and the major research projects on smart grids are faced and reported.

Citation: R. Miceli, S. Favuzza and F. Genduso, "A Perspective on the Future of Distribution: Smart Grids, State of the Art, Benefits and Research Plans," Energy and Power Engineering, Vol. 5 No. 1, 2013, pp. 36-42. doi: 10.4236/epe.2013.51005.

Keywords: Distributed Generation; Smart Grids

INTRODUCTION

Actually, the electrical distribution systems, overall Medium Voltage (MV) ones, and their design and operation criteria are subjected to deep changes, due to different factors. Among these, the main ones concern: the energy market liberalization, the new and complex energy governance policies, the environmental pollution reduction and sustainable development, the renewable energies development, the increase of energy efficiency, the costs reduction, the growth of the penetration of the so called Distributed Generation (DG). In particular, the forthcoming presence of DG in electrical distribution systems has strongly modified the nature of such systems; these systems, in fact, having, today, a radial topology and managed in a passive way (i.e. supplying energy from electrical power plants to end-users), are destined to reach an active role by means of the implementation of the typical functions of load management, demand side management, demand response and generation curtailment [1,2].

However the DG penetration determines some technical problems in electrical systems that must be faced and solved rapidly to exploit the potential benefits of DG and to really start the revision process aiming at the implementation of the so called smart grids [3,4].

The smart electrical distribution grids represent the needed evolution of the actual networks by means of a deeper implementation of automation functions, and a high level of Information and Communications Technologies (ICT) applications in order to increase the power quality and ancillary services, guaranteeing the security and economic/energetic efficiency in electric energy supplying [5].

After the analysis of the current scenario, characterized by the management and the problems solution determined from the above mentioned DG presence, the paper deals with the description of different possible models of smart grids (micro grids, active grids, local areas) putting in evidence the benefits and the necessary innovations which will be achieved in various fields involved in this important and fundamental evolution. Finally, to underline the importance of this issue, a brief presentation of the most important international research projects in this field are reported.

DISTRIBUTED GENERATION POSSIBLE BENEFITS AND PROBLEMATICS

The distributed generation can introduce in electrical distribution systems some potential benefits, such as:

- Flexibility and electrical load management [6];
- Coverage of the local load peaks;
- Diversification of the energy resources supply;
- Larger possibility of exploitation of renewable energy resources in favorable locations;
- Time deferral of investments aimed at the maintenance of transmission and distribution systems, at the building of new power plants and at the reinforcement of existing power plants and electrical systems;
- Electrical energy losses reduction [7,8].

On the contrary, the DG presence causes some technical problems that must be rapidly faced and solved like:

- The increase of short circuit currents;
- The increased complexity of automation and protection systems;
- The increased complexity of voltage regulation due to a modification of power flows;
- The unwanted MV systems islanding [9,10].

Increase of Short Circuit Currents

DG connection to distribution systems by means of synchronous and asynchronous generators causes a significant increase of short circuit currents (respect to passive systems). This issue can determine the overcomeing of dimensioning and sizing limits of circuit breakers and lines. The consequence is the necessity to substitute some conductors to adapt them to the bigger thermic solicitations and some circuit breakers to adapt them to higher breaking capacity.

Increased Complexity of Automation and Protection Systems

DG can determine the power flow inversion in some networks branches. This depends from type of DG, power size, connection points, and loads, with important consequences on protection selectivity. In fact, it is possible to have unwanted interventions over un-faulted lines, see **Figure 1**, due to faults in adjacent lines connected to the same bus bar (loss of selectivity). Other problem concerns possible non-functioning of opening and re-closing temporized systems to eliminate transient and semipermanent faults or to isolate faulted lines with the aim of limiting the time of out of service.

So it is necessary:

- The redefinition and re-setting of protection systems;
- The verification of protection selectivity taking in account of intensity, versus and during time of fault currents;
- The verification of time interventions of protection devices with the aim of generators stability.

Increased Complexity of Voltage Regulation Due to a Modification of Power Flows

As known, without DG, the voltage regulation is based on loads and passive systems features: so the voltage profile is easily determinable with regard to radial structures and stable loads (unidirectional power flows—see **Figure 2**). In presence of DG, the voltage regulation is more complex, because it depends on the sites, the sizes, the dispatching and the features of DG, as shown in Figure 3. A possible solution is a voltage regulation system based on a coordinated control of under load tap changers of HV/MV transformers and reactive power flows in distribution feeders.

Unwanted MV Systems Islanding

Today, in many countries, the MV systems islanding is prohibited, although it should constitute a significant potential advantage of DG. But islanding determines problems in the areas of safety, control and management. The safety aspects are related to permit repairing and maintaining interventions in safety conditions (electrical system out of voltage).

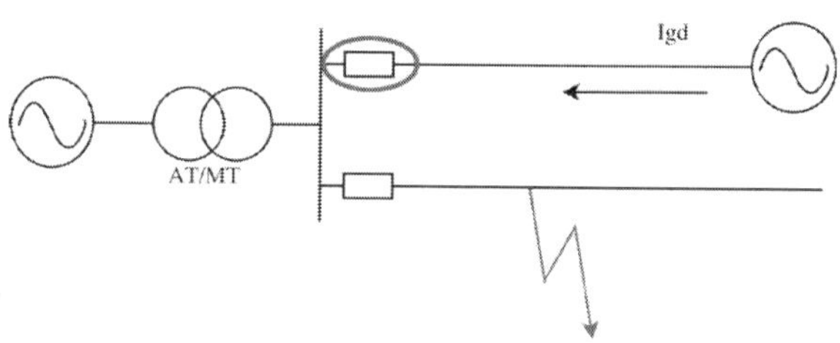

Figure 1. Loss of selectivity: unwanted intervention of a circuit breaker in an un-faulted line.

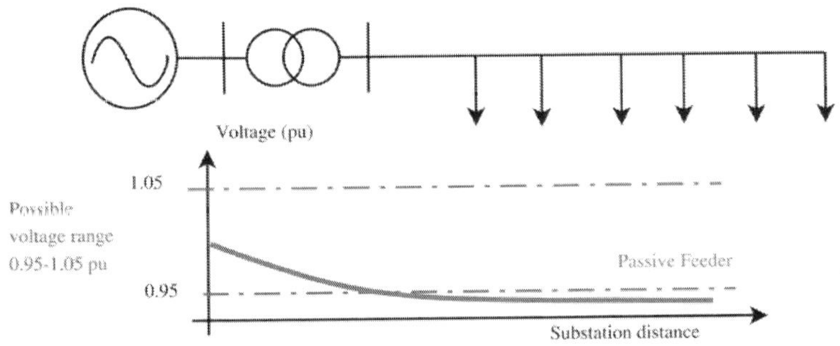

Figure 2. Voltage profile in a passive feeder.

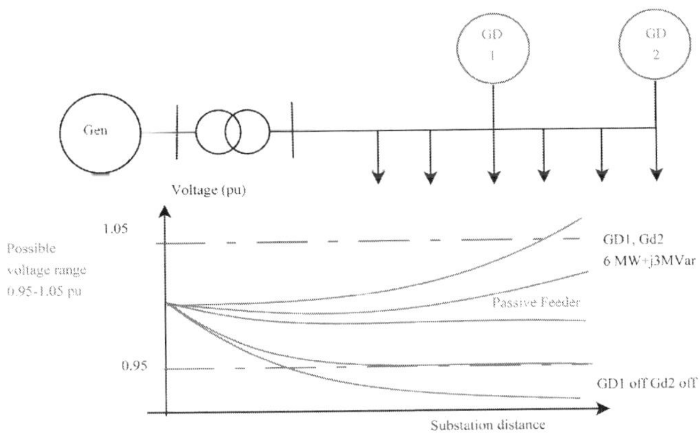

Figure 3. Different voltage profiles in an active feeder.

The control and management issues concern:

- The parallel connection of islanded system and electrical power system due to automatic opening and re-closing operation to eliminate transient and semipermanent faults;
- The ground arc extinction, with the consequence of unsuccessful of re-closing operation and the power quality (frequency and voltage variations [9,10].

CURRENT SCENARIO AND FUTURE EVOLUTION OF DISTRIBUTION SYSTEMS: THE SMART GRIDS

The classical scenario of electrical distribution systems, being passive termination of the whole electrical system, is characterized by radial topology, vertically integration with centralized generation, dispatch and control, unidirectional power flows, "connect and forget" logic for the loads, multi directional power flows in presence of DG.

The future scenario foresees an active system that guarantees connectivity through an increasing level of interaction with the consumers and meeting (at least in the future) the end users energy demands in terms of flexibility, economy and reliability, using, at the same time, the benefits of the energy market liberalization. So, it will be characterized by:

- Interconnected and meshed topology;
- Distributed logic;
- Full DG integration [11];
- Multi-directional power flows;
- Logic of integration of the loads taking large flexibility.

The major features concern:

- Larger reliability through the implementation of all the most advanced distribution automation functions (for example integrated Volt/VAR control, outage management, reconfiguration);
- Possibility to integrate the consumers and their behavior within the design and management of the network through the Demand Side Management (DSM) [12];
- Adoption of advanced communication technologies and automated controls, emergency and market demand response;
- Self-healing i.e. the possibility to detect, analyze and solve problems;
- Use of different technologies for energy generation and storage;
- Full usage of the opportunities offered by the electrical market.

To realize these goals, it is necessary to apply in electrical distribution systems an advanced distribution automation and specifically SCADA (Supervisory, Control and Data Acquisition) systems, developing and implementing:

- A monitoring of the electrical systems through sensors;
- A data transmission system (optical fibbers, PLC, GSM, WI-FI, etc.);
- A decision system and a network automation and remote control.

In this way it will be possible to realize the new electrical smart distribution grids. A smart grid is, in fact, an electrical system able to smartly integrate activities of all connected users—energy producers, consumers, prosumers—with the aim of distributing energy in an efficient, sustainable, reliable and economical favorable way [7].

The most important goal of the smart grids is to transform the functionality of the present electricity transmission and distribution grids so to provide a more useroriented service, enabling the achievement of the 20/20/20 targets and guaranteeing, in a competitive market environment, high security, power quality and economic efficiency of electricity supplying. But how and in which way the transition from actual scenario to the new one will proceed? It is hard question to answer.

At this moment it is possible to imagine for the new grids three different models, that would be, also, considered as integrated parts of smart grids, active grids, micro grids and virtual utilities [8,13].

Active Grids

An active grid is a network that does not only play the passive role of supplying the final consumers, but also in which the operator controls and/or rules the power required or generated by the loads or the generators, the bus voltages and the branch power flows. It is possible to assume an evolution in three different levels:

First-level: a simple local control of the generation at the connection point;

Second-level: a complete control system for all the distributed energy resources in the controlled area, realizing a coordinated dispatching and a voltage profile optimization (see **Figure 4**);

Third-level: creation of a strongly interconnected structure with a subdivision in cells ("local areas") responsible of their own management (protection, voltage regulation, etc.) that take part to the market, selling or buying energy to/from adjacent cells or from/to the transmission system (see **Figure 5**).

Micro Grids

A micro grid is a set of generators, loads and storage systems connected and able to operate independently from the electrical grid and that internally recreates the energy production and distribution system. In **Figure 6** is reported an example of a micro grid presenting a micro grid separation device to performa the islanding operation, an energy manager connected to several power flow controllers and protection coordinators to control and manage the energy flowing in the micro grid branches and many different type sources to inlet energy not only from the main grid, but also to allow distributed generation.

It can be considered similar to the active network cell, since it is provided with a local control system that rules the exchanges of energy among the loads, generators and external network; moreover it can stay in intentional islanding configuration, disabling the loads that accept to be part of a "load curtailment" program [1].

Virtual Utilities

A virtual utility or virtual power plant realizes an optimized management and control of a set of distributed energy resources, in which all distributed generators, loads, storage systems are coordinated taking electric market signals into account. In conclusions, in all the possible imagined scenarios, a very important role will be played by the final users. The key for the development of smart grids concerns the active demand, i.e. the possibility for consumers to actively participate as actors in the electrical system management and control.

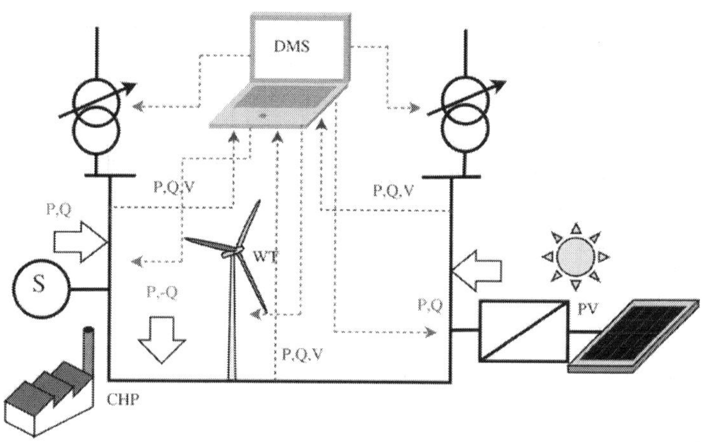

Figure 4. Decentralized control: coordinated dispatching and voltage profile optimization.

International Research Projects on Smart Grids

Many international projects and research activities are focused on smart grids, underlining the importance of this issue. The European Smart Grids Technology Platform [7] represent the strategic deployment document for Europe's electricity networks of the future. The mission is to theorize and promote a common vision about the future (2020 and over) of electric grids.

The platform presents a road map organized in six high priority areas:

1) Optimization of grids management: improvement of the cooperation between Transmission System Operators (TSO), Distribution System Operators (DSO), TSO and DSO, improvement of the grids monitoring and control;
2) Optimization of grids infrastructure: new infrastructures realization, mainly of transmission type, present infrastructures optimization, superconductor technologies development;
3) Large-scale integration of intermittent generation: development of grids to transport the energy generated by wind and photovoltaic plants, development of energy storage systems;
4) Information communication technology: development of easy, strong, safe and flexible communication infrastructures, Standardization of data types and transmission protocols;
5) Active distribution grids: spread the transmission grids services (power flow management, balance, contingencies analysis) to the distribution systems, grid topologies optimization, implementation of smart systems to the grid, communication infrastructures development;
6) New markets, users, energetic efficiency: development of new market models for the active demand integration, spread of bidirectional interface devices between users and market, knowledge boost towards the energy savings and the energetic efficiency improvement.

As known the EU Energy and climate package, that is the new European energetic strategy aiming at guaranteeing a sustainable development, electric market efficiency and quality and security of energy supply, previews some goals that must be gained in two steps: 2020 and 2050.

The first step concerns the 20-20-20 goals: reduction of 20% of greenhouse gases respect to 2005, adopted as reference year because of the start of European Emission trading (EU ETS);

20% of final consumptions of electrical energy produced by means renewables;

20% reduction of energy consumption. In 2008, to achieve these goals the European Council and Parliament have adopted the SET Plan—Strategic Energy Technology Plan, as planning and coordination document.

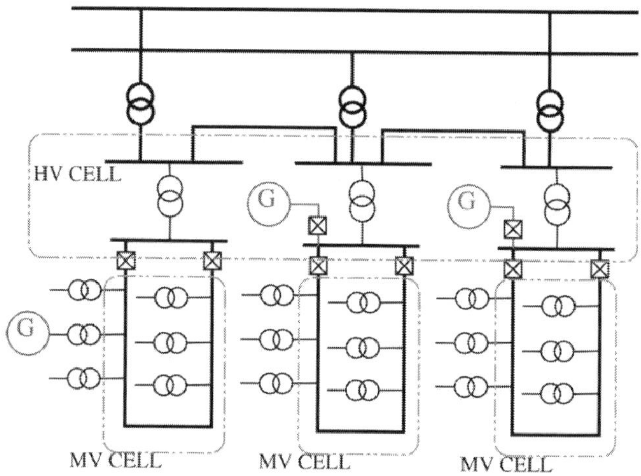

Figure 5. Distribution system organized by cells (local areas).

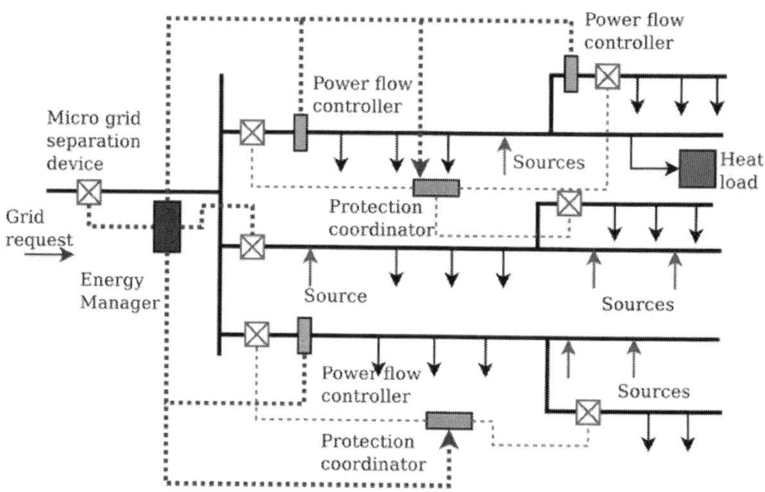

Figure 6. Example of a micro grid.

The SET Plan proposes six European Industrial Initiatives (EII), among which the smart grid. In fact, in June 2010, during the SET Plan Conference in

Madrid, the European Electricity Grids Initiative (EEGI) has been presented, with a precise research and development program. The EEGI [14] is proposed by 7 Transmission System Operators (TSO) (Amprion, Elia, Red Electrica de Espana, RTE, Tennet, Transpower, 50 Hertz) and 7 Distribution System Operators (DSO) (CEZ, EON, Enel, Erdf, Iberdrola, RWE, Vattenfall) Key partners are the European Network of Transmission Systems Operators for Electricity (ENTSO-E) and the European Distribution Systems Operators Association for Smart Grids (EDSOSG). The initiative, characterized by a time duration of 8 years (2010-2018) and by a cost of 2 billions Euro, proposes a program of research, development and demonstration (RD&D) aiming at creating a new flexible, economic, efficient, sustainable, smart electrical system.

In particular, the major features are:

- Development and integration of innovative technologies for power systems and their validation in real conditions;
- Development of suitable solutions for other energetic initiatives (solar, wind) to increase renewables and distributed energy resources (DER);
- Creation of a strictly synergy among electrical systems operators, equipment and ICT manufacturers.

In particular the EEGI vision involves electrical energy producers, end-consumers and prosumers.

One of the most important item concerns the integration of new generation and consumption models with the particular goal of creating the smart consumer, by means of: the integration, in the control and management of the electrical systems, of the active demand; the development of the energetic efficiency of end-consumers and the development of new business opportunities for electric market actors, overall for end-users.

Concerning the smart consumer, the EEGI RD&D program proposes a specific cluster named "integration of smart customers", with two projects: Active Demand Response and Energy Efficiency from integration with Smart Homes. In 2008 to support and contribute in a coordinated way to the SET PLAN goals, 14 European research institutes have created the European Energy Research Alliance (EERA). Among all the joint research program proposed by EERA, the Smart Grids one, already started, foresees 4 sub-programs (SP), each subdivided in many research activities:

- SP 1—Network Operation: Adaptation of primary control cycle and automation, network monitoring system and ICT, ancillary services, fault and outage management, distributed generation management, load management;

- SP 2—Energy Management: Simulation and Analysis Model, System Operation with DER and ICT, Market Design with DER, System Design with DER;
- SP 3—Control System Interoperability: State of the art and terminology, Classification of Control Systems, Use Cases, Technical Communication Requirements, Standards and Protocols, Cyber Security;
- SP 4—Electric Energy Storage (EES) Technologies, Performance Testing of Technologies, Integration of Storage Resources to Smart Grids possible Services, Control Algorithms for Storage, Applications in Smart Grids, Economic and Technical Benefits of incorporating an EES onto Network.

In the area of smart distribution grids, a very important European research project, financially supported by the 7th Framework Program, is working: ADDRESS (Active Distribution network with full integration of Demand and distributed energy RESourceS). ADDRESS is a large scale R&D project, involving 25 partners (distributors, networks, ICT and trade operators, research institutes, industries) from 11 European countries, with a cost of 15.7 M (9 M funded). The vision concerns:

1) Flexibility—active demand and power flow optimization both at local and global level;
2) Reliability—technologies development for distributed control, safety and efficient management;
3) Accessibility—commercial and standard barriers removal, fully DG integration, RES, active demand;
4) Economy and Savings—energy consumption and sustainable development, bills amounts reduction.

Particularly interesting is the Italian project Smart Distribution Network Operation (SDNO) and developed by ENEL-Distribuzione, the most important Italian distribution operator. The finality of the SDNO project regards:

1) The evaluation of the generators and electrical system features to enable high penetration of DG, the definition and realization

of new interface components and new network systems both hardware and software;

2) The testing of apparatus and systems in laboratories and in real electrical systems;

3) The promotion new national and international rules for connection criteria and electrical system control.

In more details, the project foresees the development or realization and testing of new functional requirements for automatic control protection systems, new Supervisory Control and Data Acquisition (SCADA) at HV/MV substation for MV system automatic control, DG dispatching and load control (DSM—demand side management). Another important aspect considered in the project is issued to new remote terminal units (RTU) in MV/LV substation for control and automation; new RTU for DG and load control (DSM); innovative rules for voltage regulation based on HV/MV transformers control and on DG dispatching.

CONCLUSIONS

All the hypothetical scenarios related to smart grids need evolution and development processes involving many aspects, which are today very interesting areas for studying and researching; in fact, the new challenges, that have to be faced, concern:

- Technical aspects: new criteria for electrical systems planning, design, control and management;
- Technological aspects: evolution of components, apparatus and systems (both hardware and software);
- Economical and policy-regulatory aspects: free markets, roles and responsibilities of all actors involved, connection rules, load shedding, generation curtailment, etc.;
- Social aspects: energy policies really sustainable, environmental impact reduction, energetic resources supplying diversification, to advantage renewable development, increase of power and services quality, lower costs for users (real competition).

A real cultural revolution has started, but, as already said by Albert Einstein"We can't solve problems by using the same kind of thinking we used when we created them".

ACKNOWLEDGEMENTS

Rosario Miceli, Salvatore Favuzza, Fabio Genduso (Italian Ministry of University and Research), by the "Sustainable Development and Energy Saving Laboratory" (SDESLAB) part of the UNINETLAB of the University of Palermo and financially supported by the project BeyWatch IST- 223888 funded by the European Community, web page: http://www.beywatch.eu/

REFERENCES

1. C. Cecati, F. Genduso, R. Miceli and G. R. Galluzzo, "A Suitable Control Technique for Fault-Tolerant Converters in Distributed Generation," IEEE International Symposium on Industrial Electronics (ISIE), L'Aquila 27-30 June 2011, pp. 107-112.
2. K. Knauss, C. Warren and D. Kearns, "An Innovative Approach to Smart Automation Testing at National Grid," Transmission and Distribution Conference and Exposition (T&D), 7-10 May 2012, pp. 1-8.
3. P. Chiradeja, "Benefit of Distributed Generation: A Line Loss Reduction Analysis," Transmission and Distribution Conference and Exhibition, Asia and Pacific, Bangkok, 2005, pp. 1-5.
4. D. L. Jia, X. L. Meng and X. H. Song, "Study on Technology System of Self-Healing Control in Smart Distribution Grid," 2011 International Conference on Advanced Power System Automation and Protection, Beijing, 16-20 October 2011, pp. 26-30. doi:10.1109/APAP.2011.6180379
5. A. O. Di Tommaso, S. Favuzza, F. Genduso and R. Miceli, "Development of Diagnostic Systems for the Fault Tolerant Operation of Micro-Grids," International Symposium on Power Electronics, Electrical Drives, Automation and Motion, Palermo, 14-16 June 2010, pp. 1645-1650.
6. F. Genduso, R. MIceli and G. R. Galluzzo, "Flexible Power Converters for the Fault Tolerant Operation of MicroGrids," XIX International Conference on Electrical Machines (ICEM), Palermo, 6-8 September 2010, pp. 1-6.
7. C. Eu, "European Smartgrids Technology Platform-Vision and Strategy for Europe Electricity Networks of the Future," European Commission, 6-8 September 2006.
8. M. Samotyj and B. Howe, "Creating Tomorrow's Intelligent Electric Power Delivery System," 18th International Conference and Exhibition on Electricity Distribution, Palo Alto, 6-9 June 2005, pp. 1-5.
9. K. Jennett, C. Booth and M. Lee, "Analysis of the Sympathetic Tripping Problem for Networks with High Penetrations of Distributed Generation," International Conference on Advanced Power System Automation and Protection (APAP), Glasgow, 16-20 October 2011, pp. 384- 389. doi:10.1109/APAP.2011.6180432

10. M. Hagh, N. Ghadimi, F. Hashemi and S. Zerbadst, "New Islanding Detection Algorithm for Wind Turbine," 10th International Conference on Environment and Electrical Engineering (EEEIC), Ahar, 8-11 May 2011, pp. 1-5.
11. A. O. Di Tommaso, F. Genduso, G. R. Galluzzo and R. Miceli, "Computer Aided Optimization via Simulation Tools of Energy Generation Systems with Universal Small Wind Turbines," 3rd IEEE International Symposium on Power Electronics for Distributed Generation Systems (PEDG), 25-28 June 2012, pp. 570-577. doi:10.1109/PEDG.2012.6254059
12. F. Aalamifar, H. Hassanein and G. Takahara, "Viability of Powerline Communication for the Smart Grid," 26th Biennial Symposium on Communications (QBSC), Kingston, 28-29 May 2012, pp. 19-23. doi: 10.1109/QBSC.2012.6221343
13. M. Loddo, "Pianificazione e Gestione Delle Reti Attive," Ph.D. Thesis, University of Cagliari, Cagliari, 2008.
14. ENTSO-E, "Roadmap 2010-2018 and Detailed Implementation Plan 2010-12," European Commission, The European Electricity Grid Initiative (EEGI), 2010.

7

A Framework for Qualifying and Evaluating Smart Grids Approaches: Focus on Multi-Agent Technologies

Gillian Basso, Nicolas Gaud, Franck Gechter, Vincent Hilaire, Fabrice Lauri

IRTES-SET, UTBM, Belfort Cedex, France.

ABSTRACT

New needs and emerging societal constraints have put the emphasis on the inadequacy of the actual electrical grid. Indeed, it is impossible, or at least very hard, to 1) integrate renewable energy sources at a great scale within the actual electric grid, 2) enable communications between the various power suppliers and consumers, 3) design several different services that meet the needs of a wide range of end users. A key solution to these issues consists in using Smart Grids (SG). SG can efficiently control power flows by means of Information Technology (IT). Technically, a SG consists of a power system and a bidirectional communication system. Multi-Agent Systems (MAS) constitute a possible technology that can be applied to

Citation: G. Basso, N. Gaud, F. Gechter, V. Hilaire and F. Lauri, "A Framework for Qualifying and Evaluating Smart Grids Approaches: Focus on Multi-Agent Technologies," Smart Grid and Renewable Energy, Vol. 4 No. 4, 2013, pp. 333-347. doi: 10.4236/sgre.2013.44040.

control and monitor the operation of power grids. Moreover, MAS exhibit distribution, adaptive and intelligent features. The goal of this paper is to propose a framework of qualification and evaluation for comparison SG approaches. First, a set of features of importance for smart grids definition is identified. Then, in a second step, some criteria are given to evaluate the impact of SG on the society. Finally, these features are applied to existing MAS approaches addressing SG in order to understand and compare their different contributions.

Keywords: Smart Grids; Qualification and Evaluation; Multi-Agent Systems

INTRODUCTION

Nowadays, a paradigm shift occurs within energy systems [1]. New needs and constraints emerge, such as: reduced carbon dioxide emissions, greater introduction of renewable energy, diversification of power transactions, etc. These new needs and constraints have put the emphasis on the inadequacy of the actual electrical grid. Indeed, it is impossible, or at least very hard, to 1) integrate renewable energy sources at a great scale within the actual electrical grid, 2) allow communications between the various power suppliers and consumers, 3) design several different services that meet the needs of a wide range of end users.

A key solution to these issues consists in using Smart Grids (SG). SG can efficiently control power flows by means of Information Technology (IT), integrating not only the supply side but also the demand side and all the devices that allow the distribution of power. Technically, a SG consists of a power system and a bidirectional communication system. It mainly focuses on applying IT to the distribution and customer sides.

As defined in [2] smart grid is the term commonly used to refer to an electrical grid whose operation has been transformed from a twentieth-century analog technology based on the pervasive use of digital technology for communications, monitoring (e.g. sensing), computation, and control.

Emerging smart grid technology allows for finegrained sensing and control. This enables highly flexible, efficient, and optimized operation, including full support for market-based, demand-side management and the accommodation of alternative generation sources, including sources of consumer-generated electricity. A smart grid is defined as having the following seven principal characteristics, as specified by the US Department of Energy's National Energy Technology Laboratory in its modern grid strategy [3]. A smart grid:

- enables active consumer participation
- accommodates all generation and storage options
- enables new products, services, and markets
- provides power quality for the digital economy
- optimizes asset utilization and operates efficiently
- anticipates and responds to system disturbances
- Operates resiliently against attack and natural disaster A similar smart grid vision is put forward in the document European Smart Grids Technology PlatformVision and Strategy for Europe's Electricity Networks of the Future [4].

Over the past ten years, many research projects have addressed the smart grids and the number of projects tackling smart grids increases every year.

The goal of this paper is to propose a framework of qualification and evaluation for a comparison of every approach for smart grids. The evaluation and comparison are based on the feature analysis approach [5]. First, a set of features of importance for smart grids is identified and defined. Then, in a second step, criteria is given to evaluate the impact of the grid on the society. Finally, these features are applied to existing MAS approaches addressing smart grids in order to understand and compare their different contributions. Among the heterogeneous approaches, some are based on the Multi-Agent Systems (MAS) paradigm. MAS are a good candidate to model and implement the intelligent components of a smart grid as they exhibit autonomy, reactivity, pro-activeness and collaborative capabilities [6,7]. A workshop dedicated to agent-based approaches for energy systems (ATES) has even been created as a forum of discussion for this research since 2010. That is why authors decide to compare only MAS approaches.

This paper is organized as follows: Section 2 presents an overview of an evaluation framework for smart grids, Section 3 details the qualification framework, Section 4 presents the evaluation framework and Section 5 provides a study of the main multi-agent approaches for smart grids.

EVALUATION FRAMEWORK OVERVIEW

Significant investments have been made, especially in European Union and United States, in order to develop, demonstrate and deploy smart grids technologies. Some recent works are trying to assess the costs, performance and benefits of smart grid technologies. There is currently no established standard in this area. Among the recent initiatives, we may notably mention:

- The 2nd EU-US workshop1 on Smart Grid Assessment Methodologies jointly organized by the Joint Research Centre and the Department of Energy on the 7th of November 2011 in Washington DC.
- The Evaluation Framework for Smart Grid Deployment Plans [8].
- The smart grid scorecard2 that provides a listing of the required features a product must have to be compliant with a smart grid.
- The incentives to determine reliable metrics, costs, and benefits from projects related to smart grids, like in [9] and [10].

The present article is part of this effort by providing an evaluation framework. The study focus on multi-agent approaches for smart grid. Though we decided to study multi-agent approaches for expertise reasons, it is worth noting that the proposed evaluation framework could also be used for approaches belonging to any other paradigms.

In the following two sections, we define a framework that decomposes the analysis and the comparison of smart grids into two steps. The first step aims at positioning the evaluated approaches and understand them. This step is divided into two dimensions:

Structural dimension represents the physical infrastructure of the smart grid. Family problem dimension defines the different classes of problems solved by the smart grid.

The second step, aims at evaluating quantitatively and qualitatively the different approaches by assessing the societal impacts of these approaches. The proposed criteria for evaluating the societal impacts are defined according to the main challenges of smart grids[3]:

Greenhouse gas reduction: How to reduce the carbon footprint of the overall supply chain? (addressed by Environmental dimension of the proposed framework)

Energy security: How to increase the network's capacity in order to manage a potentially diverse set of new requirements? How to provide interruption-free reliable power accommodating all generation and storage options, especially wind and solar power? (addressed by Structural and QoS dimensions)

Economic competitiveness & affordability: How to cost-effectively transition to a low-carbon energy system, increasing affordability? How to enable the new products, services, and markets through a flexible market to provide cost-benefit trade-offs to consumers and market participants? (addressed by Economic dimension)

Human integration: How to enable the consumers to actively participate by providing them with the choices and the incentives to modify their electricity purchasing patterns and behaviors (Smart metering, Advanced Metering Infrastructure)? (addressed by Human dimension)

For each of these challenges a set of indicators is proposed. It is not always possible to test each approach with every indicator or to obtain resources from the literature that provide results of such analysis.

QUALIFICATION FRAMEWORK DEFINITION

In the following, the different aspects of the framework are defined. These aspects are composed by the first (qualification) and second (evaluation) steps. The first step is detailed through two viewpoints or dimensions (structural and family problems) and the second step is detailed through the societal dimension. A list of criteria (written in bold face) is presented thereafter for each dimension (structural and family problem) with the corresponding definitions and explanations.

Structural Dimension—Industry Focus

The structural dimension defines the energy infrastructure of the smart grid. The energy infrastructure consists of two highly-interrelated and complex systems: the energy physical network and the energy market [11]. These two systems and their relationships are precisely described by the ontology in [12].

First, we briefly explain the main concepts of the electric power delivery system described in the ontology.

Second we highlight questions must be answered for completely describing a power grid.

- The producers are responsible for generating energy.
- The transmission system operators are responsible for operating the transport of energy on the high-voltage interconnected system, called the transmission network, in order the ensure long-term services, like the balance of supply and demand, for instance.
- The distributed system operators are responsible for operating the transport of energy on medium-voltage and low-voltage energy systems in given areas. They ensure the connections of the medium-voltage and low-voltage energy networks, called the distribution networks, to the high-voltage distribution network. They also ensure the energy delivery to customers, but without including supply.
- The suppliers are responsible for the sale of energy to customers. In a liberalized energy infrastructure, a supplier can be a wholesale customer who purchases a commodity (for instance energy) with the purpose to sell it subsequently.
- The households (or customers) purchase energy for their own use.

Physical Is the network for physical delivery taken into account?

Size What is the physical network size (network voltages, number of devices, power capacity)?

The size of a network is an intricate concept and cannot be defined precisely by a single value. For completeness, a network size must contain the different voltage of the network, and the number of devices present in the network.

Storages Which types of storage systems are used (total energy, power capacity, power variation)?

Sources Which types of sources are used (renewable, capacity factor, other characteristics defined by the ontology)?

Loads Which types of loads are used (predictable, disconnectable, other characteristics defined by the ontology)?

Scalability Is the physical network scalable in realtime?

Communication Is communication between devices possible?

The new energy infrastructures have to offer a communication between distributed appliances (for example, for a demand response [13]), that is they could allow to take into account priorities on load demands or could increase the balance between consumption and production.

Microgrids Is the physical network composed of microgrids?

Microgrids [14] can be defined as parts of the energy network that comprise intermittent sources.

This approach allows for local control of distributed generation, thereby reducing or eliminating the need for central dispatch.

Islanded Is the physical network islanded?

Network type Which type of physical network is used?

The proposed approaches are validated on the basis of experiments carried out on a simulated physical network, a real network or a combination of both.

Models If simulated, which models are used for the appliances?

Timescales If simulated, which timescales are taken into account in the energy network?

Managing energy in this system implies to take into account several strong constraints.

Besides, supply and demand occurring within the physical network must always be balanced on a short timescale (milliseconds or seconds).

Commodity trade Is the network for commodity trade taken into account?

Dynamic prices Are the prices allocated dynamically between actors?

Time frames Which timescales are taken into account in the energy market?

Depending on the possible control actions, the energy balance has to be maintained at the different time frame [15]. The wholesale market and the retail market ensure energy balance at medium time frame (minutes or hours) and long time frame (days, months or years). The balancing market ensures energy balance at small time frame (milliseconds or seconds) and medium timescale.

Family Problem Dimension—Domain Focus

As stated in [16] smart grids raise many significant challenges for the AI field. Indeed, concepts and techniques will be needed to solve the numerous problems that are not solved today by the current electricity grid. Among these problems, one can cite: the maintenance of stable voltages and frequencies, supply reliability, Distributed Energy Renewables (DER) management, heterogeneous and distributed actors, self-healing networks, ···

Follows some already well-identified problems.

Unit commitment: The problem of unit commitment (UC) aims at finding the dispatch of available generation sources that meet the electrical load with the minimum cost. In other words, the aim is to determine the combination of available generating sources or units and to schedule their respective outputs to satisfy the forecasted demand with the minimum total production cost under the operating constraints enforced by the system [17].

Demand side management: Energy demand management, also known as demand-side management (DSM) or Load Management [18], is the modification of consumer demand for energy through various methods such as financial incentives and education. Usually, the goal of demand-side management is to encourage the consumer to use less energy during peak hours, or to move the time of energy use to off-peak times such as nighttime and weekends. Peak demand management does not necessarily decrease total energy consumption, but could be expected to reduce the need for investments in networks and/or power plants by reducing demand peaks.

Demand Response: The demand response (DR) is an extension to DSM problem. The difference lies in that demand response mechanisms respond to explicit requests to shut off, whereas DSM passively shut off when stress in the grid is sensed.

Supply and demand matching: Supply and Demand Matching (SDM) is concerned with optimally using the possibilities of electricity production and consumption devices to alter their operation in order to increase the overall match between electricity production and consumption.

Vehicle to Grid The idea behind the Vehicle-to-Grid (V2G) concept, is to use the flow of power (both in input and output) of an electric vehicle. These flows can be real add-on for the electrical power grid. Indeed, these vehicles, either electric cars (BEVs) or plug-in hybrids (PHEVs), have the capability to produce AC electricity. The challenge raised by this idea is to provide electricity precisely when needed and recharge these vehicles efficiently. By communicating with the electrical power grid BEVs and PHEVs could then implement a Demand Response service.

Virtual Power Plant A Virtual Power Plant (VPP) is a cluster of distributed generation devices (such as microCHP, PV, wind-turbines, small hydro, etc.) which aggregate themselves to sell electricity. The goal of VPPs is to maximize the value for both the end user and the distribution utility by using software systems. They are dynamic, deliver value in real time, and can react quickly to changing customer load conditions. A VPP matches up a variety of distributed energy systems with intelligent demand response capabilities and aggregates those resources into an asset that acts like a centralized power plant. VPPs are similar to microgrids; however, while microgrids are very local in scope, VPPs can theoretically be deployed on a wide scale at the utility level with the ability to tap resources in real time, and with enough granularity, to control the load profiles of customers, aggregate these resources, and put them up on a trader's desk.

Self-Healing Network One of the major advantages of smart grids is to allow self-healing of the network without the intervention of technicians. This will ensure more reliable supply of electricity, and reduced vulnerability to natural disasters or attack.

EVALUATION FRAMEWORK

The points described in 3 present the upstream work for smart grid analysis. However, the objectives of the smart grids are to help companies and individuals to solve the problems due to the daily energy production and management. In order to evaluate the impact of the incoming smart grids, concepts must be defined, we integrate the concept of sustainability usually considered as a composition of Environment, Economy, and Society [19] and we also add a human dimension representing system's control and supervision capabilities.

We tried to collect them within an assessment framework combining these 4 different perspectives grouped under the umbrella term societal dimension.

The Environmental Approach

The most significant environmental impact of power generation is in the form of emission to air, ground and water. Of those emissions, the most significant in terms of impact are emission to air of carbon dioxide (CO_2)

, nitrogen oxides (NO_x), sulfur dioxide (SO_x), methane (CH_4), nitrous oxide (N_2O) and particulates. The impact on water is more complicated, involving heat, volume and emissions. The objective of this part of our framework mainly consists in determining the benefits of a smart grid's model to reduce its carbon footprint. However, the objective can be extended to nuclear wastes or other discharges influencing environment. To reduce significant environmental impact, two main goals are easily focused:

- Reduce pollutants' emissions and wastes' production (reducing oil usage, reducing nuclear power plants, etc.).
- Lower transmission and distribution losses

The Global Warming Potential (GWP)

It is therefore, important to have indicators to evaluate the performance of a smart grid in terms of environmental impact (e.g. [20, Chapter 11]). To estimate the greenhouse gases' emission impact, the United Nations Framework Convention on Climate Change[4] (UNFCCC) sets up a cost measure, based on the environment impact of the CO_2, the Global Warming Potential (GWP) [21]. With this cost measure, one can easily evaluate the greenhouse effect of every generator on the environment, given the amount of emitted gases during production, and thus it will be easier to compare them.

The difficulty of the usage of this cost measure remains in evaluating the amount of greenhouse gases' emission, but some research has been made in this problem, and it is currently possible to find the quantity of CO_2 emitted per kWh (as [22] or in [23]).

The Economic Criteria

As an economic aspect is already included in an energy network (see 3.1), it is important to define all costs, such as investments (CAPEX), ongoing costs (OPEX) but also the revenues implied by the investments to establish the interest of the problem to be resolved.

All costs and revenues are depending on the problem to be solved. As described in 3.1, the energy market contains 3 main kinds of actors with different goals. The producer side is the well-known part of the economic approach. The pertinent elements for this actor are the cost operation and the investments of energy generators. The distribution of the energy created by the new devices also implies a cost which has to be taken into account.

Adding an intelligence into a grid implies potentially a communication infrastructure. It can be interesting to manage the energy loss during the transportation. For the consumer side, the main aspects are subscription, taxes and a price of the kWh. The first and the second can vary depending on the location of the consumers, while the third also varies during time.

In [24], the authors precisely describe costs and revenues of the electricity market.

The Asia Pacific Energy Research Centre [25] provides some indicators to assess of energy efficiency from an economic perspective. They propose economic value based energy-efficiency indicators measuring the quantity of energy consumed relative to the economic/onetary value of the activity generated, denominated in a currency-related unit. This is an extension of the study published by the French Agency for Environment and Energy Management (ADEME) [26], who suggests three alternatives for comprehensively reviewing trends in energy efficiency at a sectoral or sub-sectoral level:

- ***Energy Intensity.*** Considers whether energy, as an input to production, is used efficiently. Energy intensity's analysis is generally based on relative comparisons with established benchmarks, historical trends, or other comparable energy intensities.
- ***Techno-Economic Ratios.*** Calculate, from an engineering perspective, the economic production associated with the unit or specific consumption of energy.
- ***Energy Savings Indicators.*** Endeavor to measure energy savings achieved by consumers over a period. These "techno-economic effects" essentially analyze changes in the techno-economic ratios.

In [27], authors provide a step-by-step assessment framework based on the work performed by the Electric Power Research Institute (EPRI) for conducting costbenefit analyzes of smart grid projects.

The Quality of Service Criteria

Nowadays, it would be impossible to imagine daily life without having a continuous access to energy. Offering this service to a growing number of persons and needs is very challenging. For instance, the transportation of the energy must meet numerous constraints in order to ensure power quality.

Among these constraints, one can cite: nonzero frequency impedance, harmonic variations, etc.

Some statistical indicators of transmission and distributions circuits exist [28,29] for assessing the quality of such services, among the best known, we can mention:

1) **MAIFI:** Momentary Average Interruption Frequency Index measures the weighted average number of outages that last less than ten minutes, which occurred in a year and with reference to the total connected load.
2) **ASIDI:** Average System Interruption Duration Index measures the weighted average number of outages equal to or more than ten minutes, which occurred in a year and with reference to the total connected load.
3) **ASIFI:** Average System Interruption Frequency Index measures the weighted average number of outages equal to or more than ten minutes, which occurred in a year and with reference to the total connected load.
4) **SAIDI:** System Average Interruption Duration Index is the average outage duration for each customer served.
5) **SAIFI:** System Average Interruption Frequency Index is the average number of interruptions that a customer would experience.
6) **CAIDI:** Customer Average Interruption Duration Index (SAIDI divided by SAIFI).
7) **SISI:** System Interruption Severity Index measures the ratio of the unserved energy to the system peak load.
8) **FOT:** Frequency of Trippings per 100ckt-km measures the number of forced line outages (transient and permanent or sustain) initiated by automatic tripping of relay.
9) **FLC:** Frequency Limit Compliance refers to the percentage of time during the rating period that the system frequency is within the allowable range of 60 ± 0.3 Hz.
10) **VLC:** Voltage Limit Compliance refers to the percentage of the number of voltage measurements during the rating period that the voltage variance did not exceed ±5% of the nominal voltage.

Furthermore, some standards exist such as [30], which offers practice developed out of an increasing awareness of the difficulty in comparing results obtained by researchers. The Council of European Energy Regulators provides an extensive analysis on the quality of energy supply throughout Europe [31]. It provides a collection of indicators monitoring continuity of supply (interruptions), voltage quality and commercial quality used to develop a complete benchmark.

The Human Integration Criteria

From a consumer's point of view, power grid remains a black box, an unlimited amount of energy. Unfortunately, unlike fossil-fuel or nuclear based power generation, most renewable-energy sources depend upon generally unpredictable environmental factors (solar, wind, etc.). Thus, it does not suffice to replace coal or nuclear plants with solar power plants to ensure a reliable and stable energy production as consumers are now accustomed. The advent of smart grid will also require the change in consumer behavior.

This change in behavior especially requires new interface design and new ways to present information about energy usage to the user. Numerous studies have shown the impact of how to present information about energy usage impacts consumer behavior [32-37]. It appears that human behavior with respect to energy cannot be modeled only in terms of cost-effectiveness. It is directly influenced by the following factors [32]:

- Personalized information according to a user specific configuration.
- Goals and commitments.
- Constant feedback, particularly with measures of progress toward goals.
- Financial Incentives to encourage consumers to participate Supervisory Control and Data Acquisition (SCADA) are a first step towards a monitoring interface that could participate in user behavior change.

IEC PC 118[5] is working on a standard covering the architecture of smart grid user interface, function and performance requirements of demand side systems, information exchange interfaces among demand side systems/ equipment, with the aim to support applications, such as a demand response.

Summary Statement

This evaluation Framework gives 4 criteria to evaluate the impact of smart grid on the society. In this section, we will describe numerical criteria to evaluate and compare approaches solving the same problem in the same structure. The criteria defined below are not the only to evaluate a grid but they represent the improvement of the future grid as defined in the section 2.

- The Global Warming Potential (GWP) appears to be the best existing factor to evaluate the environmental aspect of a smart grid.
- The economic factor is depending of the vision of the future grid, but all aspects rely on the price of the energy. The money (dollar, euro, etc.) is thus the obvious factor to compare several grids.
- The Quality of Service (QoS) has two perspectives, the first is the feeling of the end-users about the grid (with MAIFI, CAIDI, etc.) and the second is the network stability (with FLC, VLC, etc.). These perspectives are linked, as the first can be seen as the result of the second.
- The Human Integration (HI) is the hardest criterion to quantify. The simplest way to evaluate the human interaction with the grid is the number of settings that users can modify.

Smart grids of the same domain (see Section 3) can be evaluated quantitatively and qualitatively following these criteria to offer a comparison of the result.

MAS APPROACHES

In this section, we conduct a comparative study of the main multi-agent approaches for smart grids using the qualification and evaluation framework previously defined. After over ten years of research, indisputable progress has been made. Indeed, recent approaches are more and more efficient but the benchmark results show that many efforts are still required to offer a truly comprehensive toolbox of approaches for smart grids. The qualification of the framework can globally be set even if the structural dimension is not completely defined. On the other side, the evaluation framework cannot be used with the current analysis of the studied technologies because of the lack of numerical data. This is one of the point highlighted by this paper.

GridAgent

Developed by Infotility, the Grid Agent [38,39] Enterprise Agent Manager (EAM) Suite is the main user interface which provides centralized management and works in conjunction with a suite of pre-configured agent types, specialized editors, applications, and reporting alerting tools, including: An out-of-the-box, complete agent management collection included specialized property editors: RateEditor, ModelEditor, and RuleEditor, PlanningEditor Preconfigured visual information and analytic tools include: EventManager, Smart Dashboard, Resource Viewer, SmartFilter, Cost Manager.

The GridAgent framework implements several types of agents: "Sense and Control" and "Resource" Agents who have analytic methods to calculate optimal response based on pricing signals. "Planning and Optimization Agents manage Distributed Energy Resources (DER) devices under various operational scenarios such as optimal microgrid control strategies. "Blackboard" Agents can store databases (from several media, like Internet or the MAS).

GridAgent also offers human interactions and network protection.

This Suite is developed for managing distributed energy resources and can be used for large-scale integration of distributed energy and renewable-energy resources into real distribution systems.

Its evaluation following the given criteria are in table 1.

Homebots

Homebots [40,41] is an approach that deals with the smart distributed equipment management in a house. It is based on a multi-agent system, the agents of which are directly linked to one specific equipment. The management process is based on a computational market, which can be considered as a multi-agent systems sub-field, involving specific agents.

Table 1. Grid agent evaluation.

Structural Dimension	
Physical	Yes
Size	The system is created to manage energy network supplying 24,000 network transformers in 63 secondary networks. It has to manage the 3G/SOF including the DER, consider substation load transfer, application of HVDC, higher distribution voltage (27 kV, 33 kV, 33+kV), minimal secondary voltage, company-deployed mobile DER assets
Storage	smart charging of PHEV
Sources	Dispatch of distributed and renewable generation
Loads	Loads present in buildings like home or company (e.g. HVAC)
Scalability	Yes, agents are "plug and play"
Communication	Yes, agents play roles, which are the ability to send and receive set of messages. FIPA compliance, could use IP
Microgrids	Yes, the system enables the ability to create on-the-fly "aggregate" blocks of capacity for presentation to the energy markets.
Islanded	Yes, the system enables the ability to create intelligent Islanded grid
Network type	Real, an example is presented in [39]
Commodity trade	Yes
Dynamic prices	Yes, the demand response algorithm is based on market transactions

Time frames	unknown
	Family Problem Dimension
Demand Response	Aggregation of capacity (*i.e.* consumption). Ability to shift or curtail load or dispatch local generation. Assume that each agent has a planning execution cycle.
	Societal Evaluation Criteria
Environment	Reduce energy consumption during peak thanks to load shifting and load shedding and also use Renewables.
Economic	Each agent in the system attempts to choose among available alternatives according to the criterion of minimum operating cost
QoS	Using DER, which has the potential to provide reactive power (generated close to loads) to maintain grid voltage stability
Human	expected to operate over a large range of hardware platforms and sites (such as PDA)

In this context, every load is represented by an interactive load agent, the preference and needs of which are translated into a utility function. The agents are grouped taking into account the topology of both the electrical grid and the communication system.

The utility function is embodied by a utility interface agent. This helps to provide an interface between the utility and the load management system even if the time scales for the different kind of agents (load and utility) are not necessarily the same. In the market management paradigm, the utility interface agents can be considered as a local utility salesperson. This utility agent can thus be directly manipulated in order to make the system go in on a specific way. The utility function is computed taking into account several elements such as: load model, load's current state, utility contract, user model, expected evolution of the local market.

The Homebots evaluation following the defined criteria is made in table 2.

IDAPS

IDAPS [42] is a distributed smart grid concept proposed by Advanced Research Institute of Virginia Tech. The agents in the IDAPS MAS work in collaboration to detect upstream outages and react accordingly to allow the microgrid to operate autonomously in an islanded mode. The proposed MAS consists of:

- A control agent who monitors the system voltage, detects problems and sends signals to the main circuit.

Table 2. Homebots evaluation.

Structural Dimension	
Physical	No
Size	From one house to several houses
Storage	Not defined.
Sources	Not defined, the only source for the system is the main grid
Loads	House classical loads (wash machine, lights, etc.). Each is represented by one agent who decides to connect the load or not.
Scalability	yes, depending on the loads used.
Communication	The communication is made indirectly through the market exchanges.
Microgrids	Indirectly yes but not really mentioned
Islanded	No
Network type	In [41], both are mentioned but only simulations are presented.
Models	everything is integrated into utility function.
Timescales	Not detailed (expressed in time units)
Commodity trade	Yes
Dynamic prices	Yes, Homebots is based on market computational market economies.
Time frames	The utility function includes both local agent timescale and market timescale in the prediction that has to be taken into account.
Family Problem Dimension	
Adaptation to power production	Two examples of adaptation to power shortage are presented. In the first, the adaptation is made thanks to utility function. In the second, the adaptation is made by influencing the energy price.
Societal Evaluation Criteria	
Environment	Not available.
Economic	Minimizing energy cost based on market evolutions.
QoS	Not taken into account
Human	Human interaction can be made indirectly by influencing the market

- A DER agent who is responsible for storing associated DER information and monitoring and controlling DER power levels.
- A user agent who acts as a customer (user or load) gateway.
- A database agent who is responsible for storing system information.

IDAPS is realized with Zeus [43] multi-agent system platform, which is FIPA-compliant.

This work aims at demonstrating a practical implementation of multi-agent systems in a smart grid located at a distribution level. It also demonstrates that the agent's capability can be considered as a software alternative to a traditional hardware-based zonal protection system for isolating a microgrid. IDAPS separates the multi-agent system (developed with Zeus) and the microgrid hardware (developed with Matlab/Simulink).

The evaluation of IDAPS is presented in table 3.

IDEaS Project

The IDEaS Project has been focusing on the following problem families since its advent:

- Adaptive Home Heating ([45])
- Demand Prediction ([46,47])
- Demand-Side Management ([48,49])
- Electric Vehicles ([50-52])
- Energy Exchange ([53-55])

Table 3. IDAPS evaluation.

	Structural Dimension
Physical	Yes
Size	Circuit 9 of Virginia Tech: a residential distribution circuit at or below 12.47 kV, 34 laterals with 117 transformers serving 780 customers.
Storage	Battery Model from Matlab-Simulink 2009 [44]
Sources	Solar photovoltaic, wind turbines, microturbines, fuel cells
Loads	Typical residential needs: Lighting, Freezing, Refrigeration, Clothes Drying, Cooking, Water Heating, Space Cooling
Scalability	Yes
Communication	Yes, TCP/IP or Internet Protocol
Microgrids	One microgrid
Islanded	No, but it could be in emergency situations
Network type	Both
Models	Mathematical Formulae, Equations
Timescales	Milliseconds and hours
Commodity trade	Yes but adopting the "supply-driven-demand" approach instead of the traditional demand-driven-supply approach.
Dynamic prices	Yes
Time frames	Minute, Hour and Day. Electricity price ($/kWh) can vary from every minute to every hour;
	Family Problem Dimension
Supply-Driven-Demand Management	An IDAPS microgrid employs a bulletin board that represents an electricity market place for electricity buyers and sellers to transact business in. However, instead of matching the supply with the demand and producing market clearing prices, end-use customers make a decision, whether to buy electricity or not for their deferrable demand according to the real-time electricity pricing information offered by suppliers in the grid.
	Societal Evaluation Criteria
Environment	Not available
Economic	Not available
QoS	No real indicators, just follow grid's voltage and frequency. Response time within about half an electrical cycle (*i.e.* less than 0.008 second for a 60-Hz system) from detecting the fault to stabilizing the grid
Human	Done thanks to their "User Agent" that makes features of an IDAPS microgrid accessible to users. A user agent monitors voltage, current, active and reactive power consumption at each critical and non-critical load. It also allows users to control the status of loads based on priority pre-defined by a user. It retrieves real-time information and displays all relevant information on an electronic console on the end-user premises. This may include price ($/kWhr), quantity (kWhr) and duration (clock time) of electricity available for purchase, as well as electricity pricing information.

- Exposure in Auction Markets ([56])
- Micro-storage ([57,58])
- Virtual Power Plants ([59])

As examples, only two articles proposed within the IDEaS Project will be evaluated below using our framework. The first article tackles a problem of demand-side management. The second article deals with virtual power plants.

Demand-Side Management: A decentralized demandside management model is proposed and evaluated in [48]. Such a model aims at optimizing the deferment of loads so as to maximize the comfort in the home and minimize energy cost. The proposed method can be evaluated on the basis of the criteria shown in **Table 4**.

Virtual Power Plants: Although the production of energy with Distributed Energy Resources could potentially reduce reliance on conventional power plants, they lack the capacity, flexibility and controllability to participate in a cost-efficient way for demand supply, both in the physical electricity network and in the electricity market. The creation of Virtual Power Plants (VPP), that is entities that manage a set of DERs, has been suggested in recent years to cope with the previous quoted drawbacks. In [59], the emergence of cooperatives of VPP composed of small-to-medium size renewable electricity producers is controlled with a game-theoretic approach using a pricing mechanism and a scheme allocating payments within the cooperatives. **Table 5** evaluates the proposed technique using our framework.

Table 4. IDEaS project evaluation: demand-side management.

	Structural Dimension
Physical	Yes
Size	Composed of 5000 homes, using load profiles of 26 M homes in the UK.
Storage	Not required for this kind of problem.
Sources	Not required for this kind of problem.
Loads	Deferrable loads, including shiftable static loads (like washing machines and dishwashers) and thermal loads (like fridges, boilers and radiators) and non-shiftable static loads (lighting and cooking).
Scalability	It has been shown in the article that the proposed approach can scale well
Communication	Surely but not described.
Microgrids	Represented by *smart homes*.
Islanded	Irrelevant for this kind of problem.
Network type	Simulated.
Models	Mathematical models are used for the loads
Timescales	One decision per day and per agent.
Commodity trade	Demand-side management and pricing are dynamical and influence each other.

Dynamic prices	They are allocated dynamically, using the Real-Time Pricing (RTP) based on the macro-model of the UK electricity market, but not between actors.
Time frames	Market prices are predicted for every next day over 100 experiments.
	Family Problem Dimension
Decentralized Demand-side Management	The proposed approach consists in coordinating large populations of agents. Each agent controls and adapts the deferrable loads of a *smart home* so as to maximize the comfort and minimize energy costs. The adaptive mechanism embedded in every agent is composed of two processes. First, a Widrow-Hoff learning mechanism is used to gradually adapt the deferment of the loads based on predicted market prices for the next day. Second, a stochastic mechanism is used to reoptimize the thermal load profiles on any particular day with a probability.
	Societal Evaluation Criteria
Environment	Minimizing carbon emissions.
Economic	Minimizing energy cost.
Quality of Service	Not taken into account.
Human Integration	Selecting which loads can be deferred.

Table 5. IDEaS project evaluation: virtual plants.

	Structural Dimension
Physical	Yes.
Size	The network has a nominal capacity of 17.5 MW and can supply 12000 homes.
Storage	Not mentioned.
Sources	24 simulated wind turbines are used with profiles coming from real data of the Sotavento experimental wind farm in Spain.
Loads	Not available.
Scalability	No.
Communication	No.
Microgrids	Yes: the microgrids are represented by the cooperatives of VPP.
Islanded	Yes: only the wind turbines supply the demand.
Network type	Simulated.
Models	Analytical models are used.
Timescales	Slots of hours are used.
Commodity trade	Yes.
Dynamic prices	The prices are allocated dynamically between the cooperatives of VPP and the distribution network.
Time frames	Per hour.
	Family Problem Dimension
Virtual Power Plants	Virtual Power Plants represent collections of Distributed-Energy Resources (like storage systems or renewable-energy capacitor) that can be used more efficiently to meet load demand within an electricity supply network. Several techniques based on game theory are proposed in this article for forming and maintaining cooperative VPPs and for allowing them to reliably communicate their production to the electricity grid.
	Societal Evaluation Criteria
Environment	Not taken into account.
Economic	Not taken into account.
Quality of Service	Not taken into account.
Human Integration	Not taken into account.

PowerMatcher

The PowerMatcher [11] is one of the outcomes of the Smart House/Smart Grid European FP7 project[6]. It consists in a coordination mechanism that aims the balancing of demand and supply in (possibly multi-level) microgrids integrating Distributed Energy Resources. These microgrids are defined as clusters of sources and loads. The PowerMatcher mechanism implements supply and demand matching (SDM) using a multi-agent system and a market-based control approach. The PowerMatcher technology can be the basis of a Virtual Power Plant (VPP). The qualifying and evaluation of PowerMatcher is presented in table 6.

Within a PowerMatcher cluster, the agents are organized into a logical tree. The leafs of this tree are a number of local device agents and, optionally, a unique objective agent. The root of the tree is formed by the auctioneer agent, a unique agent who handles the price forming, i.e., the search for the equilibrium price.

Follows a Description of the three Pilots

- Netherlands Twelve have a combined heat and power (CHP) plants with high efficient 1 kW boilers running on natural gas. The other 13 have a hybrid heat pump that combines an air-to-water heat pump with condensing boilers. All 25 houses have smart meters from Itron. Each house has twelve square meters of PV paneling for a total capacity of 1400 W peak. Ten houses have a smart washing machine and dishwasher from Miele. One house has a plugin hybrid Toyota Prius car and two others each has an all-electric Volkswagen Golf car. One house has a standard lead-acid battery to store solar energy for later use.

Table 6. Power matcher evaluation.

Structural Dimension	
Physical	Yes
Size	Mass application (Hoogkerk, the Netherlands), mix of domestic clusters and selected grid segments (Mannheim, Germany), reaction to critical situations (Meltemi, Greece).
Storage	Electricity storage is present in the form of batteries
Sources	Stochastic operation devices are present in the form of photovoltaic (PV) solar cells and small-scale wind turbines; classical sources are also present depending on the field test
Loads	Deferrable loads, Shiftable static loads and classical loads depending on the field pilot.
Scalability	Yes it is one of the goals of the PowerMatcher.
Communication	The communication is implemented by a smart metering approach and via web services.
Microgrids	Yes the inherent structure defined by the PowerMatcher is hierarchical and allows to decompose a grid into smaller interacting grids in a VPP fashion.
Islanded	Yes, it was one of the constraints of the third field test.
Network type	Both simulated and real network were used
Models	The appliances are simulated according to energy profiles drawn from well-known sources
Timescales	The simulator timescale/frequency is one second.
Commodity trade	Is the network for commodity trade taken into account? Yes/No

Dynamic prices	The prices are allocated dynamically, using the Real-Time Pricing (RTP) based on the macro-model of the UK electricity market, but not between actors.
Time frames	Market prices are predicted for every next day
Family Problem Dimension	
SDM	The basic mechanism underlying PowerMatcher consists in answering the Supply and Demand Matching problem. The logical structure used also allows to tackle the VPP management problem.
Societal Evaluation Criteria	
Environment	Concerning the societal viewpoint RES and batteries have been used in pilots. Some initial results show that the use of PowerMatcher may bring from 14% to 21% of CO_2 reduction.
Economic	The peak reduction has been quantified at 4.49% (annual average) for the Greek pilot.
QoS	There is no evidence of the Quality of service issue in the available documents.
Human	The clients are integrated as real actors via the prosumer concept. Indeed, each house of the Netherlands pilot has a Home Energy Management (HEM) box, which implements the local energy management strategy of the house.

- Germany The installation of the smart meters has been done during spring 2010. The installation of first system prototypes at the end customers' premises have first started at "very friendly user's" homes. Afterwards, all BEMIs will be rolled out during October and November, so that the field trial operations can start. Meltemi comprises 220 cottages, which are fully inhabited in the summer (from May to September) and mostly empty in winter. A typical cottage in the camp is a single floor building of 70 m^2 surface. Most of the cottages are more than 30 years old.
- Greece The whole camp is supplied by a 3-phase medium-tolow voltage substation. The maximum load consumption of the site is approximately 220 kW. Furthermore, a 40 kW diesel generator is present. Finally, PV systems are installed in some of the houses as well in the entrance of the camping. The total installed capacity is 6 kWp.

CONCLUSIONS

This paper presents a framework for qualifying and evaluating smart grids approaches. This contribution is mainly based on a comparative study that is conducted based on an assessment framework analyzing each approach. The comparison is only possible for approaches that have same structure and same objective, thus the paper first provides focus on structural dimension and problem dimension to bring together approaches, then gives societal evaluation criteria in order to compare these different approaches.

This study aims at defining a cartography of the existing contributions to smart grid and analyze their strengths and weaknesses. Our objective was not to determine which approach is the best among the chosen ones. Such a choice would be dependent on many conditions specific to the deployment context. However, the survey presented in this paper should help a stakeholder with the comparison of the defined features. The reader should be aware of the following limits of this work:

- The results presented in this work for all chosen approaches are based on the available documentation (articles, technical reports and presentations). We may have missed some elements and could not be absolutely sure that the presented information is complete.
- The study is theoretical. There was no real experimentation to test the different approaches.
- The features defined for the analysis and comparison were influenced by the current state of the art of the domain which is still in its infancy. Future works may improve these features and go further.

Gillian Basso, Nicolas Gaud, Franck Gechter, Vincent Hilaire, Fabrice Lauri Future directions for this work may consist in deploying a website in order to store the presented results, enable new experiments and the addition of new features.

NOTES

[1]See http://ses.jrc.ec.europa.eu/node/69

[2]see http://www.smartgridnews.com/pdf/Smart_Grid_Scorecard.pdf

[3]Mainly according to the U.S. Department of Energy

[4]http://unfccc.int

[5]International Electrotechnical Commission, see http://www.iec.ch/dyn/www/f?p=103:7:0::::FSP_ORG_ID,FSP_LANG_ID:8701,25

[6]http://www.smarthouse-smartgrid.eu/

REFERENCES

1. M. Suga, "Nedo's Smart Grid Related Activities," ADEME-NEDO Workshop, 2009.
2. SGMM Team, "Smart Grid Maturity Model," Technical Report CMU/SEI-2011-TR-025, 2011.
3. S. Pullins, "West Virginia Smart Grid Implementation Plan," Technical Report DOE/NETL-2009/1386, West Virginia Division of Energy, National Energy Technology Laboratory, US Department of Energy, 2009.
4. "European Smart Grids Technology Platform—Vision and Strategy for Europe's Electricity Networks of the Future," Technical Report, European Union, 2006.
5. Q.-N. N. Tran and G. C. Low. "Agent-Oriented Methodologies, Chapter XII: Comparison of Ten Agent-Oriented Methodologies, Pages 341-367," Idea Group, 2005.
6. S. D. J. McArthur, E. M. Davidson, V. M. Catterson, A. L. Dimeas, N. D. Hatziargyriou, F. Ponci and T. Funabashi, "Multi-Agent Systems for Power Engineering Applications—Part I: Concepts, Approaches, and Technical Challenges," IEEE Transactions on Power Systems, Vol. 22, No. 4, 2007, pp. 1743-1752.
7. S. D. J. McArthur, E. M. Davidson, V. M. Catterson, A. L. Dimeas, N. D. Hatziargyriou, F. Ponci and T. Funabashi, "Multi-Agent Systems for Power Engineering Applications—Part II: Technologies, Standards, and Tools for Building Multi-Agent Systems," IEEE Transactions on Power Systems, Vol. 22, No. 4, 2007, pp. 17531759. doi:10.1109/TPWRS.2007.908472
8. K. Herter, T. O'Connor and L. Navarro, "A Systematic Approach for Assessing Plans to Benefit Customers and the Environment," Technical report, Herter Energy Research Solutions, Environmental Defense Fund, 2011.
9. S. J. Bossart and J. E. Bean, "Metrics and Benefits Analysis and Challenges for Smart Grid Field Projects," 2011 IEEE Energytech, Cleveland, 25-26 May 2011, pp. 1-5. doi:10.1109/EnergyTech.2011.5948539
10. M. Wakefield, "Methodological Approach for Estimating the Benefits and Costs of Smart Grid Demonstration Projects," Technical Report, Electric Power Research Institute (EPRI), 2010.
11. J. K. Kok, M. J. J. Scheepers and I. G. Kamphuis, "Intelligence in

Electricity Networks for Embedding Renewables and Distributed Generation," In R. R. Negenborn, Z. Lukszo and H. Hellendoorn, Eds., Intelligent Infrastructures, Vol. 42 of Intelligent Systems, Control and Automation: Science and Engineering, Springer, Dordrecht, 2010, pp. 179-209.

12. K. H. van Dam, "Capturing Socio-Technical Systems with Agent-Based Modeling," PhD Thesis, Delft University of Technology, Delft, 2009.
13. M. H. Albadi and E. F. El-Saadany, "Demand Response in Electricity Markets: An Overview," IEEE Power Engineering Society General Meeting, Tampa, 24-28 June 2007, pp. 1-5.
14. R. H. Lasseter and P. Paigi, "Microgrid: A Conceptual Solution," IEEE 35th Annual of Power Electronics Specialists Conference, Vol. 6, June 2004, pp. 4285-4290.
15. M. Amin, "Towards Self-Healing Energy Infrastructure Systems," IEEE Computer Applications in Power, Vol. 14, No. 1, 2001, pp. 20-28. doi:10.1109/67.893351
16. S. D. Ramchurn, P. Vytelingum, A. Rogers and N. R. Jennings, "Putting the 'Smarts' into the Smart Grid: A Grand Challenge for Artificial Intelligence," Communications of the ACM, Vol. 55, No. 4, 2012, pp. 86-97. doi:10.1145/2133806.2133825
17. S. Salam, "Unit Commitment Solution Methods," World Academy of Science, Engineering and Technology 35, 2007.
18. C. W. Gellings and J. H. Chamberlin. "Demand-Side Management: Concepts and Methods," 2008.
19. W. M. Adams, "The Future of Sustainability: Re-Thinking Environment and Development in the Twenty-First Century," Report of the IUCN Renowned Thinkers Meetings, 29-31 January 2006.
20. National Renewable Energy Laboratory, "Power Technologies Energy Data Book," US Department of Energy, 2006.
21. J. T. Houghton and Intergovernmental Panel on Climate Change, "Climate Change 1995: The Science of Climate change. Climate Change 1995 [Intergovernmental Panel on Climate Change]. Cambridge University Press, Cambridge, 1996.
22. D. Weisser, "A Guide to Life-Cycle Greenhouse Gas (GHG) Emissions from Electric Supply Technologies," Energy, Vol. 32, No. 9, 2007, pp. 1543-1559. doi:10.1016/j.energy.2007.01.008

23. International Energy Agency (IEA), "CO_2 Emissions from Fuel Combustion," OECD/IEA, Paris, 2011.
24. C. Harris, "Electricity Markets: Pricing, Structures and Economics," The Wiley Finance Series, John Wiley & Sons Incorporated, Hoboken, 2006. doi:10.1002/9781118673409
25. "Energy Efficiency Indicators, a Study of Energy Efficiency Indicators for Industry in APEC Economies," Technical Report, Asia Pacific Energy Research Centre, Institute of Energy Economics, Tokyo, 2000.
26. ADEME, "Energy Efficiency Indicators: The European Experience," ADEME Editions, Paris, 1999.
27. V. Giordano, I. Onyeji, G. Fulli, M. S. Jiménez and C. Filiou, "Guidelines for Conducting Cost-Benefit Analysis of Smart Grid Projects," JRC Reference Reports, European Commission, 2012.
28. M. Sullivan and J. Schellenberg, "Smart Grid Economics: The Cost-Benefit Analysis," Renew Grid, 2011.
29. "Performance Incentive Scheme," Technical Report, Energy Regulatory Commission, 2005.
30. "IEEE Recommended Practice for Monitoring Electric Power Quality," IEEE Std 1159-2009 (Revision of IEEE Std 1159-1995), 2009, p. 26.
31. "5th CEER Benchmarking Report on the Quality of Electricity Supply 2011," Technical Report, Council of European Energy Regulators (CEER), 2012.
32. P. M. Johnson, "Human Centered Information Integration for the Smart Grid," Technical Report CSDL-09-15, Collaborative Software Development Laboratory, University of Hawaii at Manoa, 2009.
33. T. Voolink and R. Meertens, "The Effectiveness of Electronic Feedback on Energy and Hot Water Use by Householders," Technical Report, Maastricht University, 1999.
34. J. H. van Houwelingen and W. F. van Raaij, "The Effect of Goal-Setting and Daily Electronic Feedback on InHome Energy Use," The Journal of Consumer Research, Vol. 16, No. 1, 1989, pp. 98-105. doi:10.1086/209197
35. H. Staats, P. Harland and H. Wilke, "Effecting Durable change: A Team Approach to Improve Environmental Behavior in the Household," Environment and Behavior, Vol. 36, No. 3, 2004, pp. 341-367. doi:10.1177/0013916503260163
36. A. Faruqui, S. Sergici and A. Sharif, "The Impact of Informational

Feedback on Energy Consumption: A Survey of the Experimental Evidence," Energy, Vol. 35, No. 4, 2009, pp. 1598-1608.

37. L. Becker, "Joint Effect of Feedback and Goal Setting on Performance: A Field Study of Residential Energy Conservation," Journal of Applied Psychology, Vol. 63, No. 4, 1978, pp. 428-433. doi:10.1037/0021-9010.63.4.428

38. D.A. Cohen, "GridAgents: Intelligent Agent Applications for Integration of Distributed Energy Resources within Distribution Systems," IEEE Power and Energy Society General Meeting-Conversion and Delivery of Electrical Energy in the 21st Century, Pittsburgh, 20-24 July 2008, pp. 1-5.

39. G. James, D. Cohen, R. Dodier, G. Platt and D. Palmer, "A Deployed Multi-Agent Framework for Distributed Energy Applications," Proceedings of the Fifth International Joint Conference on Autonomous Agents and MultiAgent Systems, Hakodate, 8-12 May2006, pp. 676-678.

40. J. M. Akkermans, F. Ygge and R. Gustavsson, "Homebots: Intelligent Decentralized Services for Energy Management," 4th International Symposium on the Management of Industrial and Corporate Knowledge, Rotterdam, 21-22 October 1996, pp. 128-142.

41. F. Ygge, R. Gustavsson and H. Akkermans, "Homebots: Intelligent Agents for Decentralized Load Management," DA/DSM 96 Europe Distribution Automation & Demand Side Management Vienna, Vienna, 8-10 October 1996.

42. M. Pipattanasomporn, H. Feroze and S. Rahman. "MultiAgent Systems in a Distributed Smart Grid: Design and Implementation," IEEE Power Systems Conference and Exposition, Seattle, 15-18 March 2009, pp. 1-8.

43. H. S. Nwana, D. T. Ndumu, L. C. Lee and J. C. Collis, "Zeus: A Toolkit for Building Distributed Multiagent Systems," Applied Artificial Intelligence, Vol. 13, No. 1-2, 1999, pp. 129-185. doi:10.1080/088395199117513

44. O. Tremblay, "A Generic Battery Model for the Dynamic Simulation of Hybrid Electric Vehicles," IEEE Vehicle Power and Propulsion Conference (VPPC), Arlington, 9- 12 September 2007, pp. 284-289.

45. A. Rogers, S. Maleki, S. Ghosh and N. R. Jennings, "Adaptive Home Heating Control through Gaussian Process Prediction and Mathematical

Programming," 2nd International Workshop on Agent Technology for Energy Systems (ATES 2011), Taipei, 2-3 May 2011, pp. 71-78.

46. H. Rose, A. Rogers and E. H. Gerding, "Mechanism design for Aggregated Demand Prediction in the Smart Grid," AAAI Workshop on Artificial Intelligence and Smarter Living: The Conquest of Complexity, San Francisco, 7-8 August 2011.
47. H. Rose, A. Rogers and E. H. Gerding, "A Scoring Rule-Based Mechanism for Aggregate Demand Prediction in the Smart Grid," The 11th International Conference on Autonomous Agents and Multiagent Systems (AAMAS 2012), Valencia, 4-8 June 2012, pp. 661-668.
48. S. Ramchurn, P. Vytelingum, A. Rogers and N. Jennings, "Agent-Based Control for Decentralised Demand Side Management in the Smart Grid," The 10th International Conference on Autonomous Agents and Multiagent Systems (AAMAS 2011), Taipei, 2-6 May 2011, pp. 5-12.
49. S. Ramchurn, P. Vytelingum, A. Rogers and N. Jennings, "Agent-Based Homeostatic Control for Green Energy in the Smart Grid," ACM Transactions on Intelligent Systems and Technology, Vol. 2, No. 4, 2011, p. 35.
50. E. Gerding, V. Robu, S. Stein, D. Parkes, A. Rogers and N. Jennings, "Online Mechanism Design for Electric Vehicle Charging," 10th International Joint Conference on Autonomous Agents and Multi-Agent Systems (AAMAS), Taipei, 2-6 May 2011, pp. 811-818.
51. V. Robu, S. Stein, E. Gerding, D. Parkes, A. Rogers and N. Jennings, "An Online Mechanism for Multi-Speed Electric Vehicle Charging," 2nd International Conference on Auctions, Market Mechanisms and Their Applications (AMMA), New York, 22-23 August 2011, pp. 100-112.
52. S. Stein, E. Gerding, V. Robu and N. Jennings, "A Model-Based Online Mechanism with Pre-Commitment and Its Application to Electric Vehicle Charging," 11th International Joint Conference on Autonomous Agents and Multi-Agent Systems (AAMAS), Valencia, 4-8 June 2012, pp. 669-676.
53. P. Vytelingum, S. D. Ramchurn, T. D. Voice, A. Rogers, and N. R. Jennings, "Trading Agents for the Smart Electricity Grid," The 9th International Conference on Autonomous Agents and Multiagent Systems (AAMAS), Toronto, 9-14 May 2010, pp. 897-904.

54. S. Miller, S. D Ramchurn and A. Rogers, "Optimal Decentralised Dispatch of Embedded Generation in the Smart Grid," 11th International Joint Conference on Autonomous Agents and Multi-Agent Systems (AAMAS), Valencia, 4-8 June 2012, pp. 281-288.
55. M. Alam, A. Rogers and S. Ramchurn, "A Negotiation Protocol for Multiple Interdependent Issues Negotiation over Energy Exchange," IJCAI Workshop on AI for an Intelligent Planet, Barcelona, 16-18 July 2011.
56. V. Robu, I. Vetsikas, E. Gerding and N. Jennings, "Addressing the Exposure Problem of Bidding Agents Using Flexibly Priced Options," 19th European Conference on Artificial Intelligence (ECAI), Lisbon, 16-20 August 2010, pp. 581-586.
57. "Agent-Based Micro-Storage Management for the Smart Grid," 9th International Joint Conference on Autonomous Agents and Multiagent Systems (AAMAS), Toronto, 9-14 May 2010, pp. 39-46.
58. T. Voice, P. Vytelingum, S. Ramchurn, A. Rogers and N. Jennings, "Decentralised Control of Micro-Storage in the Smart Grid," AAAI-11: 25th Conference on Artificial Intelligence, San Francisco, 7-11 August 2011, pp. 1421- 1426.
59. G. Chalkiadakis, V. Robu, R. Kota, A. Rogers and N. Jennings, "Cooperatives of Distributed Energy Resources for Efficient Virtual Power Plants," 10th International Joint Conference on Autonomous Agents and Multiagent Systems (AAMAS), Taipei, 2-6 May 2011, pp. 787-794.

8

Energy Efficiency in Smart Grid: A Prospective Study on Energy Management Systems

Hermes José Loschi, Julio Leon, Yuzo Iano, Ernesto Ruppert Filho, Fabrizzio Daibert Conte, Telmo Cardoso Lustosa, Priscila Oliveira Freitas

Department of Communications, Faculty of Electrical and Computer Engineering, University of Campinas (UNICAMP), Campinas/SP, Brazil

ABSTRACT

The term Smart Grid has become a term to represent the benefits of a smart and sophisticated electrical grid, which can meet various social expectations related to sustainability and energy efficiency. The Smart Grid promises to enable a better power management for energy utilities and consumers, to provide the ability to integrate the power grid, to support the development of micro grids, and to involve citizens in energy management with higher levels of responsibility. However, this context comes with potential pitfalls, such as vulnerabilities to cyber-security and privacy risks. In this article, a

Citation: Loschi, H., Leon, J. , Iano, Y., Filho, E. , Conte, F., Lustosa, T. and Freitas, P. (2015), "Energy Efficiency in Smart Grid: A Prospective Study on Energy Management Systems". Smart Grid and Renewable Energy, 6, 250 259. doi: 10.4236/sgre.2015.68021.

prospective study about energy management, and exploring critical issues of modeling of energy management systems in a context Smart. Grid is presented along with background of energy management systems. An analysis of the demand response condition is also presented. Finally, the advantages and disadvantages of the implementation of energy management systems, and a comparison with the Brazilian electricity system are discussed.

Keywords: Smart Grid, Management, Cloud Computing, Energy, Efficiency

INTRODUCTION

Electric energy is essential to increase productivity and ensure a high quality of life; therefore, the relationship between electric power and economic growth is crucial. However, the consequence of the current worldwide economic growth and electricity demand is the depletion of energy resources. An essential and effective way to prevent the depletion of resources and promote economic growth at the same time is the application of the concept of energy efficiency through energy management systems, this is being the basic principle of the Smart Grid [1] .

With the development of the Smart Grid, especially in the distribution grid, and with the possibility of load modeling, control over the peaks of energy demand becomes vital. The peaks of demand are serious problems and present themselves in the electrical system. The demand management in residential, commercial and industrial sectors can play an important role in reducing peak demand, reducing stress, overhead transmission and distribution lines. In many countries, there are various demand response programs, implemented for industrial and commercial loads [2] .

There are few demand response (DR) programs in use for energy management in the residential sector. Direct load restriction is the most popular method used to reduce peak demand. However, when using direct control/restriction load, the consumer comfort can be compromised. In contrast, reduction of peak demand through load displacement can benefit consumers and energy utilities [2] .

Peak demand of energy has caused adverse effects to the reliability and stability of the power system during recent decades. Reducing peak demand can reduce the risk of faults on transmission and distribution grid, consequently, the risk of interruptions. Demand response is one way to deal with peak events and avoid overload on the grid by providing the necessary flexibility through load displacement [2] [3] .

Multiple energy management systems use the concepts of demand response; however, these DR systems are not broadly implemented due to complexity of the automation involved, particularly in industrial and commercial buildings. Although the residential sector is responsible for a significant amount of electrical energy demand, few DR programs are currently used in the residential field. Direct Load Control (DLC) is the method most used by energy utilities in Brazil to manage peak demand in the residential sector, where consumer loads are adjusted in time during peak network events [4] [5] .

An energy management system that can automatically switch the operation of appliances during peak hours, can be used to management of peak demand without compromising the quality of supply to the consumer [6] .

Up recognizing the very different perspectives and priorities of individuals and organizations involved in the electrical system, the authors propose a broad analysis through a prospective study to examine the different ways for the Smart Grid meeting the demands and developments of society. This study focuses on comparing the development of "Smart Grid" in different regions of the world and demonstrating the commitment of these countries to change the social and political contexts and expectations, which often are shaped by specific regions, goals and available resources.

SMART GRID TECHNOLOGIES

The term Smart Grid means more than a single technology or even a clear set of individual technologies. Is an "umbrella" term under which various technologies of electric power systems are considered, both in hardware and software. For some people, Smart Grid is characterized primarily as the addition of an information and communication technology (ICT), superimposed in a way on existing infrastructure. For others, Smart Grid represents the installation of new transmission lines, meters, and renewable energy generation [7].

However, in order for both conditions to comply, first it is necessary to understand the legacy electric systems worldwide. The current dominant infrastructure of electric power systems involves four basic elements:

- Generation: Electric power is generated in large-scale power plants;
- Transmission: High-voltage electrical energy is transported from the plant to substations closer to consumers;
- Distribution: Low voltage energy is distributed from substations to residences and commercial buildings;
- Consumer: Electricity used for consumer devices such as refrigerators, computers, lights, pumps and other devices used by residential, commercial, and industrial devices.

The main mechanism in power generation by current systems depends on the heat produced by burning fossil fuels, division of atoms in nuclear energy, or from the hydroelectric stations water movement. Except for solar cells, almost all other forms of power generation, including the burning of fossil fuels, nuclear, biomass, hydro, wind, concentrated solar, cogeneration, and need driving a turbine to produce electricity [7] [8] .

The generation usually produces electricity with relatively low voltages ranging from 2 to 30 kilovolts (kV), depending on the size of the unit. Since electricity is generated, its tension is amplified before transmission. A critical step between the generation of electricity and long-distance transmission involves a transformer to increase the voltage. Often, the generation of electric energy occurs far from the places where the electricity is needed, making the long distances of high voltage transmission lines, a crucial part of the electrical system. The long-distance transmission voltage varies from 115 to 120 kilovolts (KV), so the transformer plays a crucial role in increasing the voltage for transmission [7] [8] .

The high-voltage transmission lines carry electricity from generating plants to local substations, where the energy is "left over" for a lower voltage and then sent through electric energy distribution networks for local users, including the industrial, commercial, and residential consumers. From the substation, the electrical energy is distributed locally within a community to individual buildings and houses. The voltage is usually reduced at the point of use, to the standard voltage of that region, which varies in different countries (with most consumers receiving 110 - 120 V in the United States and 220 - 240 V in Europe) and with the requirements of electric power use [7] [8] .

Main Technologies

The term Smart Grid represents the integration of digital technologies, sensors, and ICTs to empower and make the management on the use of electricity more reliable and efficient. Smart Grid includes technologies for the consumer (with which consumers interact) and the grid (transmission and distribution that are less visible to consumers). The Smart Grid technologies also include hardware and software [7] [8] .

One of the definitions of the term Smart Grid is the integration of various technologies, products and services, from the generation, transmission, and distribution; using advanced communication and control technologies. Figure 1 illustrates this concept.

Table 1 presents the main technologies proposals for Smart Grid and their definitions.

In addition to the above-mentioned Smart Grid technology, a more holistic approach to energy management systems is presented in the next section. A major challenge with current systems is the limited mechanisms for coordination and communication between the management of the different parts of the system. The management of the current systems' transmission and distribution are carried out through separate activities that occur in different parts of the network [8] .

ENERGY MANAGEMENT SYSTEMS IN SMART GRID

The vast majority of the energy management systems just consider the monitoring and data statistics of energy consumption of consumer electronics. For these systems, manual actuation is necessary in each device to reduce energy consumption. However, the Smart Grid technologies require management systems to be smarter and able to respond to demands related to the charge control, energy management, and timing systems with micro grids [9] .

Some architectural solutions for energy management systems are being integrated into a concept of Smart Grid; these are presented blow. Chang-Sci Choi develops an architecture using AMI solutions for an energy management system entitled EMM (Energy Monitoring and Management), interoperating this system with the Smart Grid. Among the assumptions adopted, the most important is the division of the stream of operation and settings of residences and buildings [9] .

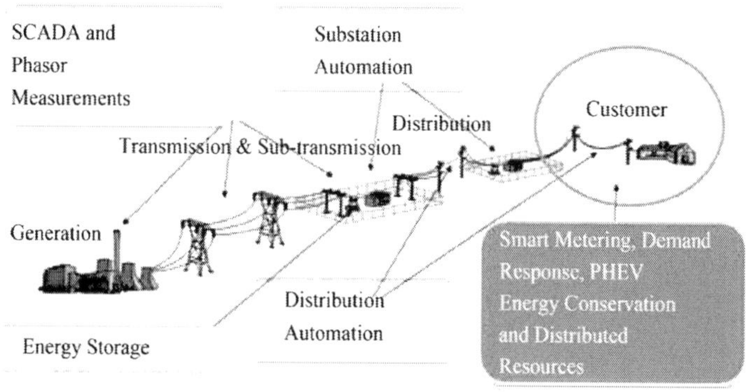

Figure 1. Concept for control and communication systems in Smart [9] .

Table 1. Main technologies proposals for Smart Grid [7].

Phases	Technologies	Definition
Generation	Inverters "Grid Tie"	Use of "Grid Tie" inverter, connecting to mains, e.g. photovoltaic systems
Transmission FACTS (Flexible Alternate Current Transmission Systems) para HVAC (High Voltage Alternating Current) e HVDC (High Voltage Direct Current) Static Deviation/Compensator VAR (Volt-amper reactive)	Synchrophasors	Measurement of Synchronized Phasor, i.e. sinusoidal measurements of AC magnitudes, and synchronized in time expressed in phasors. To determine useful information about grid performance
	AC or DC (High Voltage ou Low Voltage) voltage, transmission from generation to distribution	
	Used for static VAR compensation, have mutual inductance lines, consuming reactive power	
Test and modeling of transmission	Software for Analysis of Transmitted Power	Full suite of tools to create, configure, customize and manage power transmission systems

Transmission in cases of distribution of HVDC and generation by renewable sources	Inverters and rectifiers	Used for conversion AC-DC and DC-AC
Transmission and distribution	Substation automation	Applications of automation: Control, timing, voltage transfer, reduction and detection of loads
Transmission, distribution and substation	Relays and circuit breakers	The relay is actuated to detect any failure in the power system, disarming the breakers of the circuits, avoiding damage to the equipment
Distribution (Distribution management system-DMS)	Fault location for distribution System.	Fault locators are devices and software, usually installed in substations to identify possible fault events, calculating the distance from the point of the failure to the source monitored identified in distribution system
Distribution	AMI (Advanced metering Infraestrutura)	AMI provides the communication with the smart meters and other power management devices
Distribution, information management.	Distribution automation	The system consists of equipment's, communication infrastructure and information technology, which are used as intelligence distribution system
Information management RTU (Remote Terminal Unit) SCADA (Supervisory Control and Data Acquisition) EMS (Energy Management System)	Management system of measurement data	Used for complex data collection processes measured for multiple data recording technologies
	The function of the RTU is the remote location of the system SCADA, for monitoring and control of the necessary equipment	
	Computational system that monitors and controls infrastructure and industrial plant processes based installation	
	The EMS are tools used for operation of electrical grid, to monitor, control and optimize the performance of systems for generation, transmission and distribution	

Information management and consumer economics		Smart Meters	The smart meter registers the range of the electric energy consumption
Information management: HAN (Home are network) and LCM (Load Control Modules).	PCT (Program Communication Thermostat)	Components of the control system, to detect the temperature of a system, controlling the same in the desired set point.	
Residential Consumption	Load control receiver	Are devices used for load control, directly or indirectly, through voltage circuits, such as air conditioning thermostat.	
Consumption of the generation, distribution and transmission	Short circuit current limiter	Current limitation is the practice in electrical or electronic circuits, the imposition of an upper limit to the current that can be delivered, avoiding damage to the equipment	
Transformers generation, distribution and transmission	OLTC (On-Load Tap-Changer)	The OLTC is used to change the transformer voltage ratio, without interrupting the load service	

The sizing system EMM first considered complex apartments and residential environments popular in South Korea, with advanced infrastructure of home networks and several features, such as: electricity, gas, water, and cogeneration. Figure 2 shows the architecture that is installed via the internet providing accessibility, mobility, and interoperation with other systems. Each EMM installed performs communication with smart watches "Smart Meters", and interface with networked home appliances. To report the current status of the house, the energy measurement system sends data in real time via the Internet [9] .

For commercial buildings and industrial plants, the author proposes the BEMS (Building Energy Management System). However, these systems and its communication protocols depend on the market and corporate strategies at the time of construction, and its concepts of energy efficiency and energy management, which adopt generic and growing open communication standards and protocols, such as: BACnet, LonWorks, Modbus, KNX, WLAN, Zigbee, SNMP, IEC61850, DNP3, etc. To overcome these constraints, the author proposes the development of 2 CCL (Common Communication Layer) protocols. The first one is called BAS CCL, and the other one, EMM CCL. Figure 3 illustrates the configuration of EMM for heterogeneous integration BEMS in each building [9] .

In both systems, the server data is analyzed by software, providing information about weather and conditions of operation of the entire plant. This information can be broadly used and incorporated into a maintenance planning, among other applications [9] [10] .

Levels of data analyzed by the software can be broadly classified as:

- Level 1, Management: Supervision of computers, servers and services data management servers;
- Level 2, Automation: Smart G/W (gateway);
- Level 3, Installation: Sensors, actuators and controllers.

The software operates considering a data group, composed of service server and data management that are installed on the building control center, as illustrated in Figure 4. The Smart Gateway is installed inside the building, collecting the data of the electric power consumption through intelligent devices such as meters, sensors and actuators, etc., that are engaged in the construction of an operating system, including electricity control systems, HVAC/HVDC, lighting systems, etc. [10] .

This architecture allows the system to provide multiple functions for energy saving control, such as: control of maximum demand of energy, weather-based light and dimming control by means of sensors. The information generated by the system makes it possible to provide the function of energy efficiency through predictive analysis of the trend of energy consumption, comparing with the data in real time, through a comparative analysis with similar facilities. Another option is the provision of services, such as load shifting adapting to pricing/rates through Smart Grid function; connection with renewable energy sources; and energy exchange function. In addition, it allows a selection of ideal tariff systems appropriate to the construction standard and energy consumption of the installation [10] .

ANALYSIS OF THE CONDITION OF DEMAND RESPONSE

To meet peak demand, high-cost generating stations are required. Adding more generation was the strategy used in the past to meet the demand of electricity. Currently, the energy utilities have given more attention to demand management in order to reduce peak demand. Using the Smart

Grid concept means more than a single technology or even a clear set of individual technologies for this [2] .

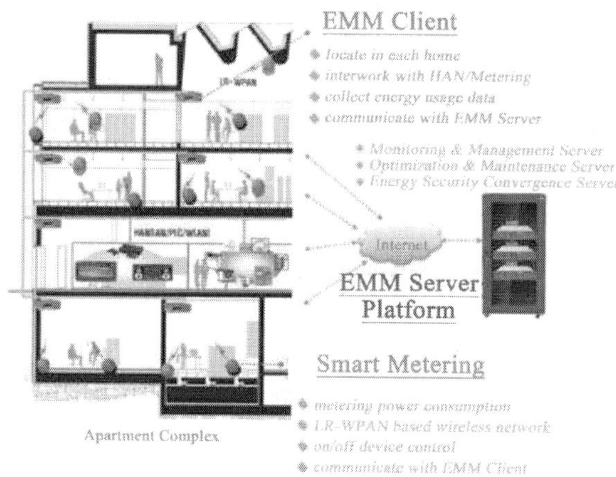

Figure 2. Systems configuration of EMM for residence [9] .

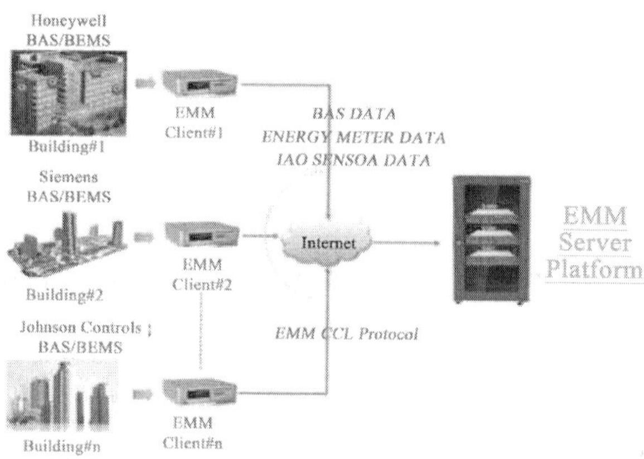

Figure 3. Systems configuration of EMM for integration with BEMS [9] .

Demand response (DR) is a key concept in energy demand management, which helps to reduce peak demand in critical situations. DR is defined as the changes in the use of electricity, for end consumers, of their normal consumption patterns in response to changes in the price of electrical energy over time, or the incentive payments intended to induce a better use of electricity at peak hours [11] .

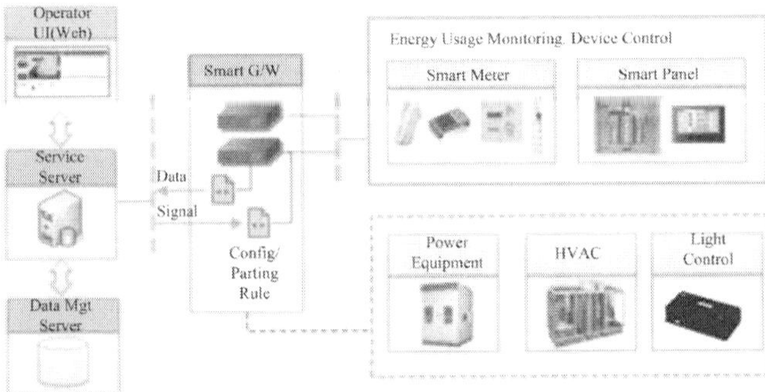

Figure 4. Performance software architecture [10].

Load management is defined as a set of objectives which aim to control directly or indirectly and/or modify patterns of electricity consumption of various consumers, aiming to reduce peak demand. This control and modification enable the supply system to meet the demand, making better use of its available generation and transmission capacity [12] .

For leveling the peaks of demand, three common strategies for load management are used: "Peak Clipping", "Load Shifting" and "Valley Filling", as illustrated in Figure 5.

"Peak Clipping": Load reduction for short peaks and periods of use, usually performed by the direct load control. In this method, the energy utilities disconnect the consumer when there is a critical situation. This direct control can be used to reduce capacity requirements, operational costs and dependence on fossil fuel generation [12] .

"Valley Filling": Creation of loads during the peak period. This helps to reduce the average price of electricity. One of the methods used in industrial production, which uses the loads generated by fossil fuels [12] .

"Load Shifting": Moves the peak loads for other periods of time without necessarily changing the global consumption. This method combines the benefit of "Peak Clipping" and "Valley Filling" moving existing loads during off-peak hours [12] .

In programs of DR, electricity consumers play an important role in the reduction of peak demand during peak hours. Consumers can move their loads and thus help the energy utilities to prevent failures and blackouts in the electrical system, reducing the probability of stress conditions of the system. Improve energy security through the DR increases productivity and customer satisfaction. The DR also eliminates the need for high-cost generators and eventually reduces the cost of electricity [2] .

In order to inform consumers with real-time data, there must be a communication link between the energy utilities and consumers. Consumers must be able to measure their electrical energy demand, in real time, in order to act for demand response events. Advanced metering infrastructure implementations (AMI) and other technologies allows the user to measure the real-time energy demand and further enhance the use of resources of DR in daily operation [13] .

Therefore, it is evident that there is a need for an automatic energy management system in DR programs, which will provide more flexibility consumers.

CLOUD COMPUTING SOLUTIONS IN SMART GRID

The term cloud computing has many definitions; in scenarios as Smart Grid, it is defined as a model for enabling convenient, on-demand network access to a shared pool of configurable computing resources (e.g. networks, servers, storage, applications, and services) that can be rapidly provisioned and released with minimal management effort or service provider interaction. The characteristics of cloud computing include on-demand service, ubiquitous network access, location independent resource pooling, rapid elasticity, and measured service.

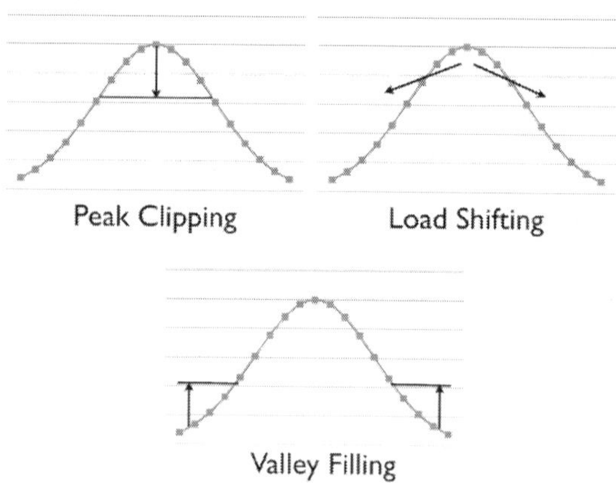

Figure 5. Strategies for load management [12] .

Cloud computing is not a universal solution. It has strengths and weaknesses, and understanding them is the key to making a decision about whether it is right for a particular application [14] . The main advantages of cloud computing are: Economy of scale: sharing of computing resources between different customers; Pay per use: customers pay for the service instead of buying software licenses and hardware; On-demand usage/flexibility: cloud services can be used almost instantly and can easily be scaled up and down; External data storage: customers' data is stored externally at the location of the cloud computing provider; Highly reliable services: clouds manage themselves in case of failures or the performance degradation [15] .

One of the most common applications for real-time data in manufacturing and process industries is SCADA, supervising remote processes over a network. With the growing popularity of cloud computing, many engineers and managers in the automation sector are looking at the possibility of using the cloud for SCADA. SCADA systems have evolved over time and have followed the progress of computing in general. As many view cloud computing as the next logical step in this evolution, enthusiastic visionaries foresee a fourth, "cloud" generation of SCADA, where an entire control system would be running in the cloud [16] . The cloud is creating a revolution in SCADA system architecture because it provides very high redundancy, virtually unlimited data storage, and worldwide data access all at a very low cost [17] .

The cloud computing can support SCADA applications in two ways: The SCADA application is running on-site, directly connected to the control network and delivering information to the cloud where it can be stored and disseminated. The control functions of the SCADA application are entirely isolated to the control network. However, the SCADA application is connected to a service in the cloud that provides visualization, reporting, and access to remote users. These applications are commonly implemented using public cloud infrastructures (PaaS cloud service). The SCADA application is running entirely in the cloud and remotely connected to the control network. The controllers are connected via WAN links to the SCADA application running entirely in the cloud. These applications are commonly implemented using private or hybrid cloud architectures (IaaS cloud service) [18] [19] .

Because of wide ranging variability of the entities in Smart Grids, there is a very high level of potential complexity in finding the optimal solution for each different Smart Grid. Smart Grid will eventually be deployed across all types of infrastructure using widespread Internet of Services, connecting all smart objects worldwide. It will become the major application domain of the Internet of Things, perhaps even referred to as the Internet of Energy [19] .

BARRIERS OF BRAZILIAN ELECTRICAL GRID

Energy management systems in residential, commercial, and industrial sectors can play an important role in reducing peak demand of electric network. Eventually, it can help in reducing overhead and stress on transmission and distribution lines. In many countries, there are various demand response programs implemented for the industrial and commercial sector. With the installation of energy management systems, it becomes possible to perform load control, mainly through the models of RTP (Real Time Pricing) and TOU (Time-of-Use) [20] .

Few demand response programs are used for management energy in the residential sector in Brazil. Direct restriction of loads is the most popular method used to reduce peak demand. But by direct control of the load, the consumer comfort can be compromised. In contrast, by the method of load displacement, the loads that have less impact to the consumer lifestyle are displaced outside of peak hours, ensuring a better stability of the network.

To analyze and consider the use of the concept of demand response is important to understand the load models "Hung" on the grid, for such function Smart Grid technologies play a fundamental role. This identification in conjunction with a proper communication between the consumer and the energy utilities and demand management of domestic charges are indispensable factors for high efficiency power management systems.

FINAL CONSIDERATIONS

Energy management systems, when developed in a context Smart Grid, have their functions enhanced, in regards of the technologies making up the system. Among the main features of an energy management system, the following stand out as the most enhanced:

- Real-time performance monitoring;
- Information to compose predictive maintenance planning;
- Energy management;

- The efficiency of the power system;
- Financial economics, avoiding any sanction from the energy utilities;
- To keep the tolerance of voltage and current to the extent allowed;
- Continuous monitoring for power quality;
- Load control and management with appropriate methodology;
- To keep consumption with the limit signed;
- Database of information that can assist future decision-making;
- Identify and correct the causes of energy disorder to avoid recurrences.

Among the main Smart Grid technologies, the most important and that directly impacts the design of an energy management system are the solutions on cloud computing. Cloud computing has established itself as an adequate means to provide resources to customers, primarily in energy management systems, with access to a large amount of information and computer storage. With cloud computing, customers do not have to manage and maintain their own information technology (IT), and are not bound to its local resources which often are limited. However, for customers and energy utilities, making sure that your cloud services are usable, an appropriate level of guarantees of Quality of Service (QoS) is needed.

In recent years, the creation of solutions in Data Center Networks (DCNS) came with rapid growth in scale and complexity, making possible hosting large applications, known as cloud hosting. This growth imposes enormous challenges to update the current datacenter infrastructure, especially considering a scenario of Smart Grid with cloud computing solutions, broadly used in energy management systems.

The proliferation of the adoption of cloud computing solutions in recent years is driven by the potential for obtaining benefits such as reduced costs, greater agility, and better use of resources. However, there are many challenges to ensure the success of these cloud-based services, and these need to be understood and managed before the major use in concepts such as Smart Grid.

However, the major current infrastructures are owned by a large number of Internet Service Providers (ISPs); and it is difficult to adopt new architectures without the agreement of all parties concerned. This includes the standardization of communication protocols and creating regulations for wide use of cloud computing solutions.

ACKNOWLEDGEMENTS

The authors would like to thank the Coordenação de Aperfeiçoamento de Pessoal de Nível Superior (CAPES), the Concelho Naional de Desenvolvimento Científico e Tecnológico (CNPq), the Fundação de Amparo à Pesquisa do Estado de São Paulo (FAPESP), the Departamento de Comunicações (DECOM), the Faculdade de Engenharia Elétrica e de Computação (FEEC), and the Universidade Estadual de Campinas (UNICAMP), for their support in the development of this research.

REFERENCES

1. Eissa, M. (2012) Energy Efficiency—The Innovative Ways for Smart Energy, the Future towards Modern Utilities.
2. Ireshika, M.A.S.T. (2014) Home Energy Management System. Universitetet I Agder, Kristiansand & Grimstad.
3. Khodaei, A., Shahidehpour, M. and Bahramirad, S. (2011) SCUC with Hourly Demand Response Considering Intertemporal Load Characteristics. IEEE Transactions on Smart Grid, 2, 564-571. http://dx.doi.org/10.1109/TSG.2011.2157181
4. Du, P. and Lu, N. (2011) Appliance Commitment for Household Load Scheduling. IEEE Transactions on Smart Grid, 2, 411-419. http://dx.doi.org/10.1109/TSG.2011.2140344
5. Gatsis, N. and Giannakis, G.B. (2012) Residential Load Control: Distributed Scheduling and Convergence with Lost AMI Messages. IEEE Transactions on Smart Grid, 3, 770-786. http://dx.doi.org/10.1109/TSG.2011.2176518
6. Li, J., Chung, J.Y., Xiao, J., Hong, J.W.-K. and Boutaba, R. (2011) On the Design and Implementation of a Home Energy Management System. Proceedings of the 6th International Symposium on Wireless and Pervasive Computing, Hong Kong, 23-25 February 2011, 1-6. http://dx.doi.org/10.1109/iswpc.2011.5751338
7. Stephens, J., Wilson, E.J. and Peterson, T.R. (2015) Smart Grid (R) Evolution. Cambridge University Press, Cambridge.
8. Jiang, T., Yu, L. and Cao, Y. (2015) Energy Management of Internet Data Centers in Smart Grid.
9. Choi, C.-S., Ian, J.I., Park, W.-K., Jeong, Y.-K. and Lee, I.-W. (2011) Proactive Energy Management System Architecture Interworing with Smart Grid. Proceedings of the IEEE 15th International Symposium on Consumer Electronics, Singapore, 14-17 June 2011, 1-4.
10. Park, K., Kim, Y., Kim, S., Kim, K., Lee, W. and Park, H. (2011) Building Energy Management System based on Smart Grid. Proceedings of the IEEE 33rd International Telecommunications Energy Conference, Amsterdam, 9-13 October 2011, 1-4.
11. Balijepalli, V.S.K.M., Pradhan, V., Khaparde, S.A. and Shereef, R.M. (2011) Review of Demand Response under Smart Grid Paradigm. Proceedings of the 2011 IEEE PES International Conference on

Innovative Smart Grid Technologies-India, Kollam, 1-3 December 2011, 236-243. http://dx.doi.org/10.1109/iset-india.2011.6145388

12. Paracha, Z.J. and Doulai, P. (1998) Load Management: Techniques and Methods in Electric Power System. Proceedings of the International Conference on Energy Management and Power Delivery, Singapore, 3-5 March 1998, 213- 217.
13. Medina, J., Muller, N. and Roytelman, I. (2010) Demand Response and Distribution Grid Operations: Opportunities and Challenges. IEEE Transactions on Smart Grid, 1, 193-198. http://dx.doi.org/10.1109/TSG.2010.2050156
14. Bashir, A.K., Ohsita, Y. and Murata, M. (2015) Abstraction Layer Based Distributed Architecture for Virtualized Data Centers. Proceedings of the Sixth International Conference on Cloud Computing, GRIDs, and Virtualization, Nice, 22-27 March 2015, 62-67.
15. Frey, S., Disch, S., Reich, C., Knahl, M. and Clarke, N. (2015) Cloud Storage Prediction with Neural Networks. Proceedings of the Sixth International Conference on Cloud Computing, GRIDs, and Virtualization, Nice, 22-27 March 2015, 68-72.
16. Mcilvride, B. (2012) Will SCADA Envolve to the Cloud? http://real-timecloud.com/
17. Combs, L. (2011) Cloud Computing for SCADA. http://www.controleng.com/single-ar
18. Conway, G., Carcary, M. and Doherty, E. (2015) A Conceptual Framework to Implement and Manage a Cloud Computing Environment. Proceedings of the Sixth International Conference on Cloud Computing, GRIDs, and Virtualization, Nice, 22-27 March 2015, 138-142.
19. Markovic, D.S., Zivkovic, D., Branovic, I., Popovic, R. and Cvetkovic, D. (2013) Smart Power Grid and Cloud Computing. Renewable & Sustainable Energy Reviews, 24, 566-577. http://dx.doi.org/10.1016/j.rser.2013.03.068
20. Albadi, M.H. and El-Saadany, E.F. (2007) Demand Response in Electricity Markets: An Overview. Proceedings of the 2007 IEEE Power Engineering Society General Meeting, Tampa, 24-28 June 2007, 1-5.

9

Voltage Control in Smart Grids: An Approach Based on Sensitivity Theory

Morris Brenna[1], Ettore De Berardinis[2], Federica Foiadelli[1], Gianluca Sapienza[3], Dario Zaninelli[1]

[1]Politecnico di Milano – Department of Energy, Milan, Italy; [2]CESI S.p.A., Milan, Italy; [3]Politecnico di Milano – Department of Energy in Collaboration with ENEL Distribuzione S.p.A., Milan, Italy.

ABSTRACT

Due to the development of Distributed Generation (DG), which is installed in Medium-Voltage Distribution Networks (MVDNs) such as generators based on renewable energy (e.g., wind energy or solar energy), voltage control is currently a very important issue. The voltage is now regulated at the MV busbars acting on the On-Load Tap Changer of the HV/MV transformer. This method does not guarantee the correct voltage value in the network nodes when the distributed generators deliver their power. In this paper an

Citation: M. Brenna, E. Berardinis, F. Foiadelli, G. Sapienza and D. Zaninelli, "Voltage Control in Smart Grids: An Approach Based on Sensitivity Theory," Journal of Electromagnetic Analysis and Applications, Vol. 2 No. 8, 2010, pp. 467-474. doi: 10.4236/jemaa.2010.28062.

approach based on Sensitivity Theory is shown, in order to control the node voltages regulating the reactive power exchanged between the network and the dispersed generators. The automatic distributed voltage regulation is a particular topic of the Smart Grids.

Keywords: Voltage Regulation, Reactive Power Injection, Distributed Generation, Smart Grids, Sensitivity Theory, Renewable Energy

INTRODUCTION

Due to the development of Distributed Generation (DG), which is installed in Medium-Voltage Distribution Networks (MVDNs) such as generators based on renewable energy (e.g., wind energy or solar energy), voltage control is currently a very important issue.

The voltage of MVDNs is now regulated acting only on the On-Load Tap Changer (OLTC) of the HV/MV transformer [1]. The OLTC control is typically based on the compound technique, and this method does not guarantee the correct voltage value in the network nodes when the generators deliver their power [2,3].

When a generator injects power in the network, the voltage tends to rise. In HV networks this phenomenon happens mainly when reactive power is injected, because the resistance is negligible if compared with the inductive reactance [4]. Instead in MVDNs the resistance is not negligible and the result is that an injection of active power also increases the voltage.

In other words the so-called Pq - QV decoupling [5], which is a typical of HV networks, is inexistent in MVDNs. The P variations are "coupled" with the voltage variations.

If no precautions are taken, in particular network conditions the overcome of the maximum admissible voltage can happen in any nodes.

When a generator injects power, the voltage rises in all network nodes, but some nodes are mainly influenced than others by the power injection. This influence can be obtained using a Sensitivity method.

In this paper an approach based on Sensitivity Theory is shown, in order to control the network voltage using the reactive power exchanged between network and the distributed generators. This approach allows to control the voltage in the long term period. Besides, fastdynamic voltage disturbances are not taken into account [6].

After the theoretical analysis, a numerical example is shown, in order to validate the proposed theory.

The proposed method differs from the others used in HV networks analysis, based on the Jacobian Matrix [1,2-4] and its application is easy.

The topological proprieties that results from the theoretical analysis imply that the proposed sensitivity method can be easily implemented in automatic voltage control devices, in order to obtain the distributed voltage regulation.

The automatic voltage regulation in a distributed manner is a typical topic of the Smart Grids context.

The paper is structured in the following way. In Section 1, the proposed voltage control method is shown, and an overview on the voltage profiles with DG, are given. In Section 3, the proposed Sensitivity approach is studied, referring to a MV test network, composed by four nodes. Finally, in Section 4, a numerical application is presented, in order to validate the proposed theory.

THE PROPOSED CRITERIA TO CONTROL THE NETWORK VOLTAGE WITH DISTRIBUTED GENERATION

Many methods can be used to control the voltage in network nodes (network voltages). The proposed method varies the reactive power exchanged between the generators and the network while maintaining the OL-TC in a fixed position for a particular load condition.

Let us suppose that the Automatic Voltage Regulator (AVR) that controls the OLTC maintains the MV bus-bar voltage at the rated value (1 p.u.), assuming that the transformer taps are adequate.

For passive grids, when no generators are connected to the MVDN, the voltage profile (VP; i.e., the voltage values along a line) decreases monotonically (see profile a in **Figure 1**) due to the load absorptions. When the generators are connected and inject power into the MVDN, the nodal voltages increase and the VP is no longer monotonic, as shown in profile b in **Figure 1** (profile b). This phenomenon also occurs if generators work at unitary power factor (i.e., only active power is injected due to the non-negligible network resistance) [7].

It is important to note that, in steady-state, the condition maintained at the MV busbar by the AVR decouples the MV feeders, and the result is that each feeder works without the influence of the other lines. In other words, the loads and generators connected to other feeders do not influence the VP of the considered line.

Typically, the generators installed in Smart Grids are based on renewable energy; therefore, their power-time profiles are unknown. Due to the high generated power and a possibly low load condition, the voltage in some nodes can thus exceed the maximum admissible value (V_{max}; i.e., the voltage threshold [8]) defined by the standards.

Of course the voltage threshold is strictly related with the settings of the voltage relays installed in the network, e.g. at the generator nodes [9].

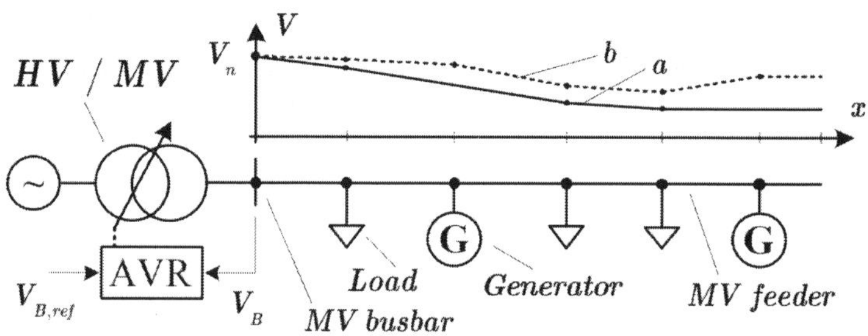

Figure 1. Voltage profiles in a MV feeder with and without Distributed Generators

If the generators are able to control the injected or absorbed reactive power, the network voltage profiles can be modified by acting on the reactive powers. It is clear that each controllable generator needs a Generator Remote Terminal Unit (GRTU) that is connected to a central control system to set the generator reactive power, (i.e., to control the exciter of the synchronous generators [1] or act on the inverter control if the generator is inverter-based) [10,11]. In this work, the central control is called the Generator Control Centre (GCC). In addition, we use a hierarchical control structure [12,13].

Let us suppose that the voltage is measured only in the generator nodes by the GRTUs. This assumption does not affect the generality of the proposed method because a Measuring Remote Terminal Unit connected to the GCC can be installed in each node that must be controlled.

When the voltage in the ith node exceeds V_{max}, the GRTU installed in the same node sends the signal Voltage Threshold Overall (VTO) to the GCC using a communication channel. The GCC then selects the generator in the jth node that has the maximum influence on the voltage of the ith node, the "Best Generator" (BG), and switches it to the reactive power absorption (RPA) mode. Therefore, the voltage in the ith node tends to decrease.

The problem is thus to determine the best generator and ensure that the GCC chooses it. In this work, a sensitivity-based method is proposed to select the BG.

Moreover, we suppose that the generators can only be switched in the RPA mode by the GCC by a constant power factor. Therefore, if P_j is the active power injectted by the generator connected to the jth node, then it absorbs the reactive power $Q_j = P_j \tan\varphi_j$ (where $\cos\varphi_j$ is the minimum power factor of the generator) when it is switched during RPA. In other words, we assume that no continuous reactive power modulation is possible.

An example of the procedure described above is shown in **Figure 2**. Let us suppose that load Ld suddenly decreases its power (for example, due to a trip) and V_2 exceed V_{max}.

The GRTUs of G2 send the signal VTO to the GCC that must choose the BG using the sensitivity method. Assuming that the BG is G1, it will be switched by the

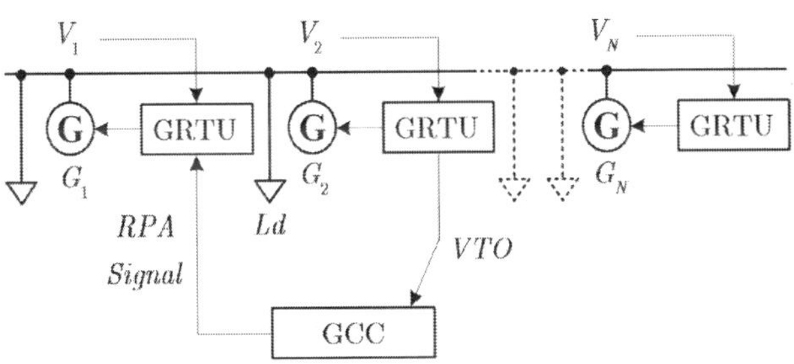

Figure 2. Voltage control using GRTU and GCC

GCC in the RPA mode; therefore, the reactive power absorbed by G1 becomes $Q_1 = P_1 \tan\varphi_1$.

As explained in the following, the GCC must know the reactive power that each controllable generator can absorb in order to choose the BG. We suppose that this information is acquired by the GCC using a polling technique on each GRTU.

THE PROPOSED SENSITIVITY APPROACH

Classical Sensitivity Theory Overview

The classical sensitivity theory used in HV network analysis to perform primary and secondary voltage regulation [14] is based on the Jacobian Matrix and reveals the relationships between the nodal voltages (magnitude and phase) and the nodal power injections (active and reactive). The relationships mentioned above are represented by the following matrix expression [2]:

$$\begin{bmatrix}[\Delta E]\\ [\Delta \vartheta]\end{bmatrix} = \begin{bmatrix}\left[\frac{\partial P}{\partial V}\right] & \left[\frac{\partial P}{\partial \vartheta}\right]\\ \left[\frac{\partial Q}{\partial V}\right] & \left[\frac{\partial Q}{\partial \vartheta}\right]\end{bmatrix}^{-1} \begin{bmatrix}[1] & [0]\\ [0] & [1]\end{bmatrix} \begin{bmatrix}[\Delta P^*]\\ [\Delta Q^*]\end{bmatrix} \tag{1}$$

where $[\Delta E]$ and $[\Delta \vartheta]$ are, respectively, the nodal voltage magnitudes (rms) and phase variations corresponding to the nodal active or reactive power injections $[\Delta P^*]$ and $[\Delta Q^*]$ ([1] is the identity matrix).

Equation (1) can be rewritten in the following compact form:

$$\begin{bmatrix}[\Delta E]\\ [\Delta \vartheta]\end{bmatrix} = [s] \begin{bmatrix}[\Delta P^*]\\ [\Delta Q^*]\end{bmatrix} \tag{2}$$

where:

$$[s] \triangleq \begin{bmatrix}\left[\frac{\partial P}{\partial V}\right] & \left[\frac{\partial P}{\partial \vartheta}\right]\\ \left[\frac{\partial Q}{\partial V}\right] & \left[\frac{\partial Q}{\partial \vartheta}\right]\end{bmatrix}^{-1} \begin{bmatrix}[1] & [0]\\ [0] & [1]\end{bmatrix} \tag{3}$$

is the (injection) sensitivity matrix. The method descries above is generally valid, but its computational complexity is too high for practical voltage analysis in MVDNs. For radial networks, only the voltage magnitude is needed to control the nodal voltages. The proposed theory is easier than classical theory, and it is suitable for radial MVDNs.

The Proposed Theory

In this section, the proposed theory for choosing the BG is outlined. The method is first described in general and considers the possibility of reactive power regulation for all nodes.

After the general treatment, the analysis focuses on a realistic network in which the reactive power can only be controlled in some nodes (generator nodes).

Let us consider the network depicted in **Figure 3**, which is a four-node test MVDN.

The general loads $Ld_1 \ldots Ld_4$ are represented using constant PQ models. Positive P (or Q) corresponds to the absorbed power by the load. Negative P (or Q) corresponds to the injected power in the network (i.e., the general load is really a generator). The per-phase equivalent circuit is shown in **Figure 4**.

The lines $L_{01} \ldots L_{24}$ are modeled using the RL-direct sequence equivalent circuit [15], but the shunt admittances are neglected. The node 0 represents the MV busbar, which is regulated at a constant voltage value E_0 by the AVR of the OLTC. This reference voltage coincides with the rated value $E_0 = V_n/\sqrt{3}$.

Because the busbar is regulated at E_0, we can characterize the generic node ı using the difference V_{o1} between the magnitude of the busbar voltage and the node voltage E_i. In other words, we can write:

$$V_{0i} = E_0 - E_i \tag{4}$$

In radial networks, (4) can be calculated as the sum of the voltage differences between adjacent nodes from the ith node toward the MV busbar. For example, if i=3 (see **Figure 4**), (4) becomes:

$$V_{03} = E_0 - E_3 \tag{5}$$

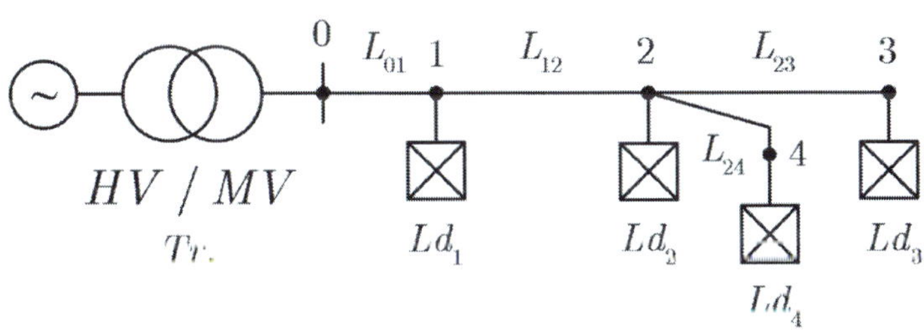

Figure 3. The considered four nodes test MVDN

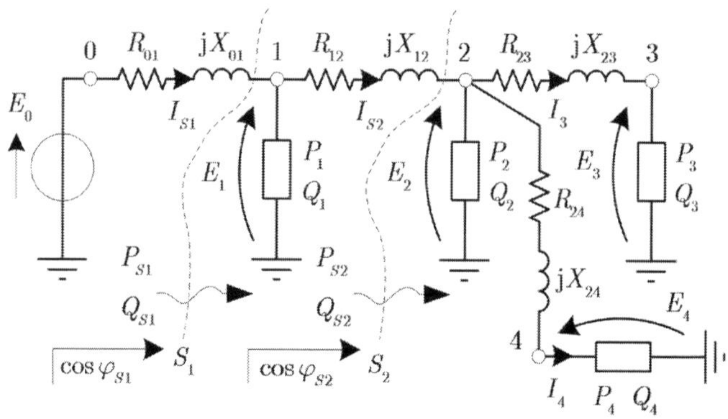

Figure 4. The per-phase equivalent circuit

By adding and subtracting E_1 and E_2 in (5), we obtain:

$$\begin{aligned} V_{03} &= (E_0 - E_1) + (E_1 - E_2) + (E_2 - E_3) \\ &= V_{01} + V_{12} + V_{23} \end{aligned} \tag{6}$$

where V_{03} is the sum of the voltage differences V_{01} , V_{12} and V_{23}.

V_{23} can be calculated considering the network parameters and the line power flows as follows:

$$\begin{aligned} V_{23} &= E_2 - E_3 \\ &= R_{23} I_3 \cos\varphi_3 + X_{23} I_3 \sin\varphi_3 \\ &= \frac{R_{23} E_3 I_3 \cos\varphi_3 + X_{23} E_3 I_3 \sin\varphi_3}{E_3} \\ &= \frac{R_{23} P_3 + X_{23} Q_3}{E_3} \end{aligned} \tag{7}$$

where $\cos\varphi_3$, P_3 and Q_3 are the power factor and the active and reactive (per-phase) powers of the load Ld_3, respectively. I_3, R_{23} and X_{23} are the current, resistance and reactance of the line L_3.

Normally, the nodal voltages are close to the rated voltage E_n. Applying this assumption to (7) leads to:

$$V_{23} \cong \frac{R_{23} P_3 + X_{23} Q_3}{E_n} \tag{8}$$

Similarly, considering nodes 1 and 2, we can write:

$$\begin{aligned} V_{12} &= E_1 - E_2 \\ &= R_{12} I_2 \cos\varphi_{S2} + X_{12} I_2 \sin\varphi_{S2} \\ &= \frac{R_{12} E_2 I_2 \cos\varphi_{S2} + X_{12} E_2 I_2 \sin\varphi_{S2}}{E_2} \\ &\cong \frac{R_{12} P_{S2} + X_{12} Q_{S2}}{E_n} \end{aligned} \quad (9)$$

where P_{S2} and Q_{S2} are the active and reactive powers through the section S_2 and $\cos\varphi_{S2}$ is the power factor for the same section. For P_{S2} and Q_{S2}, we can write:

$$P_{S2} = P_2 + P_3 + P_4 + P_{R23} + P_{R24} \quad (10)$$

$$Q_{S2} = Q_2 + Q_3 + Q_4 + Q_{X23} + Q_{X24} \quad (11)$$

where P_{R23} and P_{R24} are the power losses in R_{23} and R_{24}, while Q_{X23} and Q_{X24} are the reactive powers absorbed by X_{23} and X_{24}. These active and reactive losses are negligible compared to the load powers. Applying this assumption to (9), (10) and (11) leads to:

$$P_{S2} \cong P_2 + P_3 + P_4 \quad (12)$$

$$Q_{S2} \cong Q_2 + Q_3 + Q_4 \quad (13)$$

and:

$$V_{12} \cong \frac{R_{12}\left(P_2 + P_3 + P_4\right) + X_{12}\left(Q_2 + Q_3 + Q_4\right)}{E_n} \quad (14)$$

Finally, the voltage difference V_{01} is:

$$V_{01} = E_0 - E_1 \cong \frac{R_{01} P_{S1} + X_{01} Q_{S1}}{E_n} \quad (15)$$

where:

$$P_{S1} \cong P_1 + P_2 + P_3 + P_4 \quad (16)$$

$$Q_{S1} \cong Q_1 + Q_2 + Q_3 + Q_4 \quad (17)$$

are the powers through section S_1.

Using (6) with (15), (9) and (8), we can say that V_{03} is a function of all loads and active and reactive powers, i.e., $P_1 \ldots P_4$ and $Q_1 \ldots Q_4$. The same observation is valid for E_3:

$$E_3 = E_0 - V_{03} = E_0 - \left(V_{01} + V_{12} + V_{23}\right) \quad (18)$$

because E_0 is constant. In other words, we can write:

$$E_3 = f(P_1,...,P_4,Q_1,...,Q_4) \quad (19)$$

Equation (19) shows that an active/reactive power variation (in the general j node) that is defined as:

$$\Delta P_j = P_j^f - P_j^0 \quad (20)$$

$$\Delta Q_j = Q_j^f - Q_j^0 \quad (21)$$

where P_j^f (Q_j^f) and P_j^0 (Q_j^0) are the final and initial power values, respectively, produces a voltage variation in node 3 that is defined as:

$$\Delta E_3 = E_3^f - E_3^0 \quad (22)$$

In this treatment, we only consider the reactive power variations (i.e., $\Delta P_j = 0$) because we assume that only the reactive power can be used to control the node voltages.

The variation ΔE_3 can be calculated by linearizing (19) and considering only the reactive power variations. In particular, we can write:

$$\Delta E_3 = \frac{\partial E_3}{\partial Q_1}\Delta Q_1 + \frac{\partial E_3}{\partial Q_2}\Delta Q_2 + \\ + \frac{\partial E_3}{\partial Q_3}\Delta Q_3 + \frac{\partial E_3}{\partial Q_4}\Delta Q_4 \quad (23)$$

The terms $\partial E_i / \partial Q_j$ in (23) indicate the "gain" from the voltage variation ΔE_i in node i when a reactive power variation ΔQ_j occurs in node j. In other words, they are sensitivity terms.

According to (18), we can obtain:

$$\frac{\partial E_3}{\partial Q_2} = -\frac{X_{01} + X_{12}}{E_n}$$
$$\frac{\partial E_3}{\partial Q_3} = -\frac{X_{01} + X_{12} + X_{23}}{E_n}$$
$$\frac{\partial E_3}{\partial Q_4} = -\frac{X_{01} + X_{12}}{E_n} \quad (24)$$

Substituting equation group (24) into (23) has important implications. If we have a reactive injection in any node, i.e., $\Delta Q_j < 0$ (in this case $j = 1...4$), then $\Delta E_3 > 0$ in node 3 (i.e., the voltage increases). Then, if we were to reduce the voltage in any node, we must absorb reactive power from the network (i.e., $\Delta Q_j > 0$) by using, for example, the distributed generators.

If the above analysis that focuses on node 3 is extended to all network nodes, (23) has a general matrix relationship:

$$\begin{bmatrix} \Delta E_1 \\ \Delta E_2 \\ \Delta E_3 \\ \Delta E_4 \end{bmatrix} = \begin{bmatrix} \frac{\partial E_1}{\partial Q_1} & \frac{\partial E_1}{\partial Q_2} & \frac{\partial E_1}{\partial Q_3} & \frac{\partial E_1}{\partial Q_4} \\ \frac{\partial E_2}{\partial Q_1} & \frac{\partial E_2}{\partial Q_2} & \frac{\partial E_2}{\partial Q_3} & \frac{\partial E_2}{\partial Q_4} \\ \frac{\partial E_3}{\partial Q_1} & \frac{\partial E_3}{\partial Q_2} & \frac{\partial E_3}{\partial Q_3} & \frac{\partial E_3}{\partial Q_4} \\ \frac{\partial E_4}{\partial Q_1} & \frac{\partial E_4}{\partial Q_2} & \frac{\partial E_4}{\partial Q_3} & \frac{\partial E_4}{\partial Q_4} \end{bmatrix} \begin{bmatrix} \Delta Q_1 \\ \Delta Q_2 \\ \Delta Q_3 \\ \Delta Q_4 \end{bmatrix} \quad (25)$$

which in a compact form yields:

$$[\Delta E] = [s_Q][\Delta Q] \quad (26)$$

where $[s_Q]$ is the reactive sensitivity matrix, $[\Delta Q]$ is the reactive power-variations vector and $[\Delta E]$ is the nodal voltages vector.

Calculating the partial derivatives contained in $[s_Q]$, we have Equation (27).

$$[s_Q] = -\frac{1}{E_n} \begin{bmatrix} X_{01} & X_{01} & X_{01} & X_{01} \\ X_{01} & X_{01}+X_{12} & X_{01}+X_{12} & X_{01}+X_{12} \\ X_{01} & X_{01}+X_{12} & X_{01}+X_{12}+X_{23} & X_{01}+X_{12} \\ X_{01} & X_{01}+X_{12} & X_{01}+X_{12} & X_{01}+X_{12}+X_{24} \end{bmatrix} \quad (27)$$

After analyzing this form of (27), we can say that this matrix can be built using the following inspection rule:

"The element i, j is the arithmetic sum of the reactance of the branches in which both the powers absorbed by node i and node j flow multiplied by $-1/E_n$".

For example, in (27), the element 2, 4 is

$$-(X_{01} + X_{12})/E_n$$

because the powers delivered by node 2 and node 4 flow in branches 01 and 12.

The Choice of the Best Generator

The BG is the generator that has the greatest influence on node i, which is the node where the voltage exceeds the threshold.

Thus, after analyzing (25), we can say that the BG is the generator that maximizes the following product, which we call the "sensitivity product":

$$\frac{\partial E_i}{\partial Q_j} \Delta Q_j \quad (28)$$

For example, if the node with a voltage that exceeds V_{max} is i= 2 and the BG is connected to node j= 4, the sensitivity product $(\partial E_2/\partial Q_4)\Delta Q_4$ is the highest compared to the other products contained in row 2 of the sensitivity matrix. In addition, in order to choose the BG, it is necessary to evaluate the single products (28) of the row that represents node i. Thus, the value ΔQ_j is needed and is acquired as the GCC polls the GRTUs, as stated previously.

The procedure described above suggests a way of defining the "sensitivity table" $[T_S]$ that contains the single sensitivity products. For the MVDN represented in **Figure 4**, $[T_S]$ takes the following form

$$\begin{bmatrix} \frac{\partial E_1}{\partial Q_1}\Delta Q_1 & \frac{\partial E_1}{\partial Q_2}\Delta Q_2 & \frac{\partial E_1}{\partial Q_3}\Delta Q_3 & \frac{\partial E_1}{\partial Q_4}\Delta Q_4 \\ \frac{\partial E_2}{\partial Q_1}\Delta Q_1 & \frac{\partial E_2}{\partial Q_2}\Delta Q_2 & \frac{\partial E_2}{\partial Q_3}\Delta Q_3 & \frac{\partial E_2}{\partial Q_4}\Delta Q_4 \\ \frac{\partial E_3}{\partial Q_1}\Delta Q_1 & \frac{\partial E_3}{\partial Q_2}\Delta Q_2 & \frac{\partial E_3}{\partial Q_3}\Delta Q_3 & \frac{\partial E_3}{\partial Q_4}\Delta Q_4 \\ \frac{\partial E_4}{\partial Q_1}\Delta Q_1 & \frac{\partial E_4}{\partial Q_2}\Delta Q_2 & \frac{\partial E_4}{\partial Q_3}\Delta Q_3 & \frac{\partial E_4}{\partial Q_4}\Delta Q_4 \end{bmatrix} \quad (29)$$

Row i represents the node in which we want to control the voltage, and column j represents the nodes in which we can control the reactive power. The BG is the generator connected to node j that has the maximum absolute value of the sensitivity product in position i, j. By finding the maximum sensitivity product in row i, we automatically choose the BG because the location corresponds to column j of the maximum sensitivity product.

It is clear that, for a general network with N nodes, the sensitivity table takes the following form:

$$\begin{bmatrix} \frac{\partial E_1}{\partial Q_1}\Delta Q_1 & \frac{\partial E_1}{\partial Q_2}\Delta Q_2 & \cdots & \frac{\partial E_1}{\partial Q_N}\Delta Q_N \\ \frac{\partial E_2}{\partial Q_1}\Delta Q_1 & \cdots & \cdots & \cdots \\ \cdots & \cdots & \cdots & \cdots \\ \frac{\partial E_N}{\partial Q_1}\Delta Q_1 & \cdots & \cdots & \frac{\partial E_N}{\partial Q_N}\Delta Q_N \end{bmatrix} \quad (30)$$

It is important to note that, if it is not possible to regulate the reactive power (e.g., if in that node there is a load or a non-controllable generator) in a node j, then $\Delta Q_j = 0$ and, consequently, the sensitivity product in the position i, j of the sensitivity table is 0.

Comparing (29) with (25), we can say that each element i, j of $[T_S]$ represents the line-to-ground voltage variation in node i when a reactive power variation occurs in node j.

In the following section, a numerical example of the sensitivity method application is shown.

APPLICATION OF THE PROPOSED METHOD

The network considered in this numerical application is represented in **Figure 5**.

During normal network operation, we have four generators and eight loads. The generator and load characteristics are summarized in **Table 1** (S is the apparent power) and **Table 2**, respectively (three-phase powers are represented in these tables).

We suppose that the generators normally operate with a unitary power factor (i.e., no reactive power is injected in the nodes).

The per-kilometer reactance of the cable lines is $x = 0.17\Omega/km$, which is a typical value for Italian MVDNs. The line lengths and parameters are summarized in **Table 3**.

Let us suppose that each generator is connected to its GRTU that measures the nodal voltage and communicates with the GCC. Moreover, let us suppose that G_5 cannot regulate the reactive power because it is not designed for this purpose. The MV busbar is regulated at the rated voltage (1 p.u.), which is 20 kV (line-to-line). In this example, the voltage threshold V_{max} is 1.05 p.u.

Using load-flow software, we calculated the voltage E in the generator nodes (nodes 4, 5, 6, and 7) for normal network operation. The results are shown in **Figure 6** (Normal Operation).

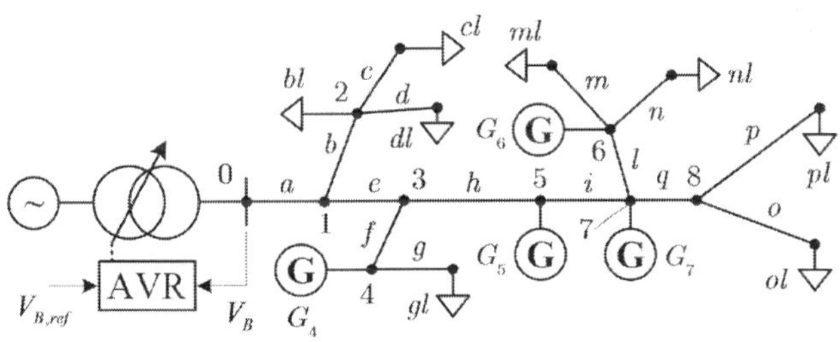

Figure 5. The network considered in the numerical application

Table 1. Loads characteristics

Load	*S [MVA]*	*cos*	*P [MW]*	*Q [MVAR]*
bl, nl	2	0.95	1.9	0.62
cl	7	0.92	6.44	2.74
dl	3	0.92	2.76	1.18
gl	2.08	0.95	1.99	0.62
ml	2.58	0.93	2.4	0.94
ol	1.98	0.96	1.9	0.57
pl	1.5	0.92	1.38	0.59

Table 2. Generators characteristics

Generator	*P [MW]*
G_4	6
G_5	1.75
G_6	4.5
G_7	3.75

Table 3. Lines parameters

Line Name	*L [km]*	*X [Ω]*
a, l, q	2	0.34
b, f, g, i, m, n	1	0.17
c, d, o, p	0.5	0.085
e	15	2.55
h	5	0.85

If line b trips (e.g., due to a fault), loads bl, cl and dl are cut off from the supply, which causes the voltage to increase in the network. In particular, if the load-flow is re-computed to take into account the new network configuration, we obtain the results shown in **Figure 6** (Tripped Line).

It is important to note that, if the voltage exceeds the maximum threshold V_{max} in node 5, the GRTU connected to G_5 sends the VTO signal to the GCC that must choose the BG using the sensitivity table.

We suppose that the three-phase reactive powers absorbable by each generator that were collected from the last poll are those summarized in **Table 4**, which also contains the corresponding power factors cosφ. To calculate the sensitivity table, we need the single-phase powers. Therefore, the reactive powers shown in **Table 4** have to be divided by three. It is important to note that the reactive powers calculated this way correspond to ΔQ_j because Q_j^0 is zero (see (21)). The ΔQ_j values are shown in **Table 5**.

The voltage exceeds the threshold in node 5. Thus, we only consider the fifth row of the sensitivity table. According to the inspection rule mentioned above, this row is as follows:

$$\begin{bmatrix} 0 & 0 & 0 & t_{S5,4} & t_{S5,5} & t_{S5,6} & t_{S5,7} & 0 \end{bmatrix} \quad (31)$$

where the single sensitivity products $t_{Si,j}$ are:

$$t_{S5,4} = -\frac{1}{E_n}(X_a + X_e)\Delta Q_4 = -164.53\ V \quad (32)$$

$$t_{S5,5} = -\frac{1}{E_n}(X_a + X_e + X_h)\Delta Q_5 = 0\ V \quad (33)$$

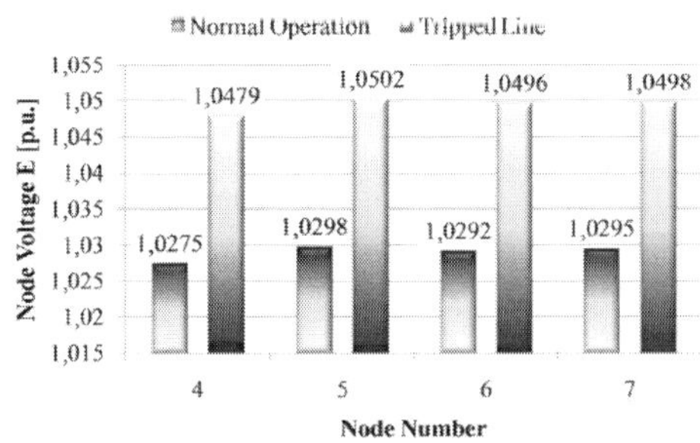

Figure 6. Load-Flow results with the Network Normal Operation

Table 4. Reactive powers absorbable by the generators

Generator	$Q[MVAR]$	$cos\varphi$
G_4	1.97	0.95
G_5	0.00	1
G_6	0.91	0.98
G_7	1.23	0.95

Table 5. Reactive Power Variations in the Generator Nodes

Generator	ΔQ_j [*MVAR*]
G_4	0.66
G_5	0.00
G_6	0.30
G_7	0.41

$$t_{S5,6} = -\frac{1}{E_n}(X_a + X_e + X_h)\Delta Q_6 = -98.65\ V \quad (34)$$

$$t_{S5,7} = -\frac{1}{E_n}(X_a + X_e + X_h)\Delta Q_7 = -133.07\ V \quad (35)$$

The maximum sensitivity product (in absolute value) corresponds to generator 4, (i.e., j= 4). Thus, the BG is G_4.

Equation (32) provides important information. If G_4 performs the considered reactive power variation, the line-to-ground voltage variation in node 5 is:

$$\Delta E_5 = -164.53\ V \rightarrow -0.0142\ p.u. \quad (36)$$

Then, considering (22) (rewritten for node 5), and $E_5^0 = 1.0502\ p.u.$ from **Figure 6**, we can say that the voltage value after the reactive power variation is:

$$E_5^f = E_5^0 + \Delta E_5 = 1.036\ p.u. \quad (37)$$

which is less than the voltage threshold V_{max}.

Equation (37) shows the theoretical result obtained using the proposed method. We checked this value using load-flow software:

$$E^f_{5\ load\ flow} = 1.034\ p.u. \quad (38)$$

The percentage error between (38) and (37) is:

$$\varepsilon_{\%} = \frac{E^f_{5\ load\ flow} - E_5^f}{E^f_{5\ load\ flow}} \cdot 100 = -0.19\ \% \quad (39)$$

which is negligible and demonstrates the validity of the proposed approach.

CONCLUSIONS

The proposed sensitivity method allows the voltage within network acting on single generators to be regulated by choosing the most effective generator on the controlled node (i.e., the Best Generator). This is a very important feature in grids that have distributed generation (e.g., in a Smart Grid context).

The proposed method uses a topological approach. Moreover, the sensitivity table can be constructed automatically.

In addition to the BG choice, the proposed method also evaluates the voltage in all network nodes after a reactive power variation.

After choosing the BG, but before its commutation during RPA, it is possible to verify that the voltage variation in the other nodes is tolerable for the connected loads. Moreover, it is necessary to verify that the threshold settings of the voltage relay installed in the same nodes.

When a generator is switched during RPA, it works with a non-unitary power factor; the reactive power flow increases along the lines and increases the power loss [16].

This phenomenon is negligible in HV networks because the line resistance is typically smaller than the line reactance, but is important to consider in MV networks.

Therefore, if network analysis reveals that the RPAswitching produces high losses, voltage control using the reactive power variation must only be used for temporary voltage variation mitigation (i.e., during emergency conditions).

Morris Brenna, Ettore De Berardinis, Federica Foiadelli, Gianluca Sapienza, Dario Zaninelli The possible future develops of this work could be focused on the optimization of the forecasted power-time profiles of the loads and generators applying both the se nsitivity approach and distributed voltage measurement.

REFERENCES

1. R. Marconato, "Electric Power Systems," Vol. 2, CEI, Milano, 2008.
2. P. Kundur, "Power System Stability and Control," McGrawHill, New York, 1994.
3. Y. Rosales Hernandez and T. Hiyama, "Distance Measure Based Rules for Voltage Regulation with Loss Reduction", Journal of Electromagnetic Analysis and Applications (JEMAA), Vol. 1, No. 2, June 2009, pp. 85-91.
4. F. Saccomanno, "Electric Power Systems" Wiley-Interscience IEEE Press, Piscataway, 2003.
5. G. Andersson, "Modeling and Analysis of Electric Power Systems" Lecture 227-0526-00, ITET ETH Zürich, Zürich, 2008.
6. Y. Abdel-Rady, I. Mohamed and E. F. El-Saadany "A Control Scheme for PWM Voltage-Source DistributedGeneration Inverters for Fast Load-Voltage Regulation and Effective Mitigation of Unbalanced Voltage Disturbances" IEEE Transactions on Industrial Electronics, Vol. 55, No. 5, May 2008, pp. 2072-2084.
7. P. M. S. Carvalho, P. F. Correia and L. A. F. M. Ferreira, "Distributed Reactive Power Generation Control for Voltage Rise Mitigation in Distribution Networks," IEEE Transactions on Power Systems, Vol. 23, No. 2, 2008, pp. 766-772.
8. P. N. Vovos, A. E. Kiprakis, A. R. Wallace and G. P. Harrison, "Centralized and Distributed Voltage Control: Impact on Distributed Generation Penetration," IEEE Transactions on Power Systems, Vol. 22, No. 1, 2007, pp. 476-483.
9. P. M. Anderson, "Power System Protection," IEEE Press, Piscataway, 1999.
10. N. Mohan, T. M. Undeland and W. P. Robbins, "Power Electronics: Converters, Applications, and Design", Wiley, 1995.
11. M. H. Rashid, "Power Electronics Handbook", Academic Press-Elsevier, 2007.
12. L. L. Grigsby, "Electric Power Generation, Transmission and Distribution," CRC Press-Taylor & Francis Group, Boca Raton, 2006.
13. F. A. Viawan and D. Karlsson, "Coordinated Voltage and Reactive Power Control in the Presence of Distributed Generation," PES

General Meeting - Conversion and Delivery of Electrical Energy in the 21st Century, IEEE, Pittsburgh, 2008, pp. 1-6.

14. S. Corsi, "Wide Area Voltage Regulation & Protection" 2009 IEEE Bucharest Power Tech Conference, Bucharest, June 28 -July 2, Bucharest, pp. 1-7.
15. A. Gandelli, S. Leva and A. P. Morando, "Topological Considerations on the Symmetrical Components Transformation", IEEE Transactions on Circuits and Systems I: Fundamental Theory and Applications, Vol. 47, No. 8, August 2000, pp. 1202-1211.
16. H. M. Ayres, L. C. P. da Silva, W. Freitas, M. C. de Almeida and V. F. da Costa, "Evaluation of the Impact of Distributed Generation on Power Losses by Using a Sensitivity-Based Method," IEEE Power & Energy Society General Meeting, Calgary, 2009, pp. 1-6.
17. A. Kishore and E. F. Hill, "Static optimization of Reactive Power Sources by use of Sensitivity Parameters", IEEE Transactions on Power Apparatus and Systems, Vol. PAS-90, No. 3, 1971, pp. 1166-1173.

SECTION 3 :
ANALYSIS AND MODELING OF RENEWABLE ENERGY

10

Reliability Evaluation of Renewable Energy Share in Power Systems

Zeyad A. Haidar, Abdullah M. Al-Shaalan

Electrical Engineering Department, King Saud University, Riyadh, Saudi Arabia.

ABSTRACT

In this research, Renewable Energy (RE) represents the existing power systems with different levels. However, because of the intermittent nature of these sources, it is necessary to analyze systems' reliability with different RE penetration levels. This work presented a simulation method for reliability evaluation of renewable penetrated power systems. Some reliability indices were proposed for the case of power systems with renewable power plants. The adopted approach used the historical data of renewable energy resources, mainly wind and solar to estimate the power that can be generated and compared with the demand to find the power mismatch. Therefore, this approach can be utilized to determine the penetration level that renewable

Citation: Haidar, Z.A. and Al-Shaalan, A.M. (2018),"Reliability Evaluation of Renewable Energy Share in Power Systems". Journal of Power and Energy Engineering, 6, 40-47. doi: 10.4236/jpee.2018.69006.

energy can be shared, and it also helps the system operators in deciding the percentage of the generation that RE power plant can provide.

Keywords Reliability Evaluation, Renewable Energy, Wind Energy, Solar Energy

INTRODUCTION

Integration of renewable energy sources in power system has the advantage of reducing CO_2 emission, hence assisting in resolving the global warming problem. It also helps decrease our dependency on fossil fuel in power sectors. Therefore, RE resources are considered a part of the solution to mitigate environmental problems caused by the use of conventional energy resources. In this sense, these resources offer clean energy generation, allowing the electrification of no-connected and remote areas, contribute to decrease dependence on fossil fuels. They also have improved their technologies and reduced costs. However, the capital cost of renewable power plant is very high compared to conventional power plants. But the operation and maintenance cost is cheaper than the one for conventional generation and will continue to decline with recent technical development.

Since most of renewable resources are intermittent in nature, it is advantageous to utilize more than one resource when available. Hybridizing of renewable resources improves the power system reliability, efficiency and economy, especially in places with good characteristics of sun and wind.

Integration of wind turbine generator and PV in conventional small isolated power system significantly lowers operating cost by offsetting costly fuel consumed by diesel generators. However, limitations in the energy available from PV system and their intermittent behavior degrade the system reliability. Therefore, cost benefit analysis associated with application of PV is incomplete without corresponding reliability assessment [1] .

The reliability of PV systems has been the subject of several research studies going back to at least the 1970's. Early work was focused on component reliability [2] [3] . Current research tends to focus on grid-tied rather than off-grid systems. The research often tries to develop new methods of conceptualizing or calculating reliability, often using probabilistic approaches. A Markov Reward Model was developed in [4] to incorporate reliability into grid-tied PV performance analysis. In [5] , the authors proposed a new technique to incorporate reliability into the levelized cost of energy of grid-tied PV systems. The impact of distributed generation

DG on the reliability of power system is investigated [6] . They studied optimization of placement of DG to improve the power system reliability using the VCS algorithm.

Methods for calculating the reliability of off-grid systems using the loss of load probability method are implemented in [7] based on probabilistic models. Off-grid systems are also considered in [8] , where the authors integrate reliability aspects into the design of such systems. Other existing research considers the reliability of hybrid systems [9] and multi-grids [10] using various probabilistic techniques.

This paper presents a method to evaluate the reliability of renewable power systems. Three scenarios were used: a power system with wind energy, solar energy and hybrid system. Although the actual case includes a variable load, the load was assumed to be constant for the sake of simplicity. The historical data of wind speed and solar radiation for Riyadh were used to calculate the generation power from wind and/or solar energy at each minute for one month. Since solar PV panels' output changes with their operating temperature, the ambient temperature was also included in the study. The aim of this study is to present or propose a method to classify the availability of renewable energy based on the historical data of the renewable resources, like wind speed and solar radiation. The availability of renewable resources can cover or feed the whole load, or it can share a percentage of the total load and the remaining is fed or supplied using conventional power plants. Therefore, this paper introduces new indices for renewable energy reliability assessment that can be used in power systems planning and operation. These indices are 25% energy index, 50% energy index, 75% energy index and 100% energy index. Each index gives the probability that the expected available renewable resources will cover 25%, 50%, 75% and 100% of the energy demand respectively during the study time.

MATHEMATICAL MODELING

The proposed indices in this study are calculated using the following equation:

$$N\%\ \text{index} = P(re > N\%)N\% \quad (1)$$

where re is the available renewable energy, $P(re>N\%)$ the probability that re is > than N% of the load, N represents 25, 50, 75 or 100 or any percent.

Therefore, 25% index gives the percent of time (over the period of study) at which the expected renewable energy is less than 25% of load, 50% index gives the percent of time at which the expected renewable energy is less than 50% of load so on.

This index attempts to answer the question: how many days (or hours) do we expect that renewable energy resources will supply N% of the load during a specified period? The benefit of using this index is to determine the adequacy of renewable resources to supply a specified portion of the load, and consequently; the conventional generation that is required to supply the remaining part of the total load. The accuracy of this method depends on the number of weather data that is included in the calculation and can be improved further, if data from forecasting models are inserted in the calculation of the renewable resources.

The wind power generated by wind turbines is calculated using the following equation [11]:

$$P_w = 0.5N_t A C_p w^3 \eta \quad (2)$$

where ρ is the air density in kg/m^3, A is the rotor swept area in m^2, C_p is the turbine coefficient of performance, w is the wind speed at hub position m/s, N_t is the number of turbines and η is the net efficiency. The theatrical value of C_p is 59.3%.

The solar energy can be calculated using the following equation [12] :

$$P_S = IA_p N_p \eta \quad (3)$$

where I is the solar radiation in w/m^2, A_p is the panel area, N_p is the number of panels, and ηη is the panel efficiency.

The simulation was carried out using the following steps:

1) For each i Î n, calculate the load and available ER (based on historical data of solar radiation and wind speed), where i is the time step (minutes or hours) and n is the total time.

2) Compute $REL = \frac{RE}{load}$ for each i, where REL is the level of RE w.r.t. load.

3) Classify the results of step (2) into groups based on level of RE penetration preset by the utility.

RESULTS AND DISCUSSIONS

Solar Energy

The number of PVs used in this study is 2000 panels. It is assumed that there are two solar PV plants at different locations and hence different solar radiations. Each location has 1000 panels of 250 W_p.

Figure 1 shows the solar PV energy that is produced during the month of study. The load power is 100 MW and is shown in the figure as horizontal line. The output power of PV panels is less than the load at some minutes and more than the load at others. However, the output power is not similar for all days. In some days the output is fluctuating, and this is indicated by the black areas in the Figure 1 like minute 250,000 (17th of the month). The peak of the solar power is not the same each day due to change in the solar radiation and ambient temperature.

Figure 2 shows the values of the proposed indices for solar energy alone. It shows that the availability of solar energy to supply 25% of the demand during the month is 0.4786. Therefore, we can expect solar PV power (with 2000 panels) to supply 25% of the load over a period equals approximately 50% of the month (15 days). Consequently, conventional energy will supply the remaining (75% of the load) during these days and the whole load on the others 15 days. However,

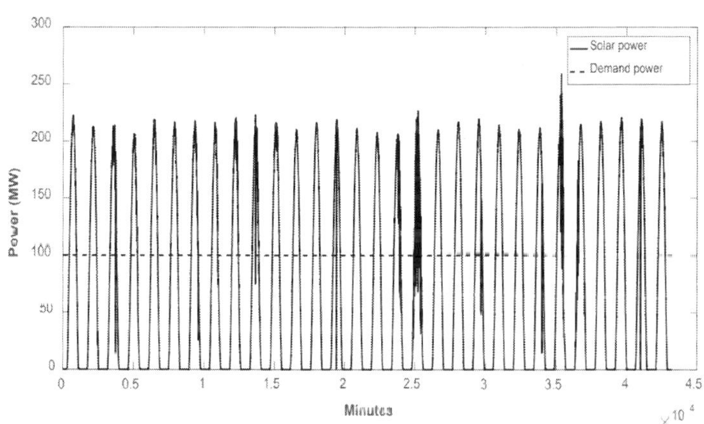

Figure 1 The PV panels' energy produced during the month minutes.

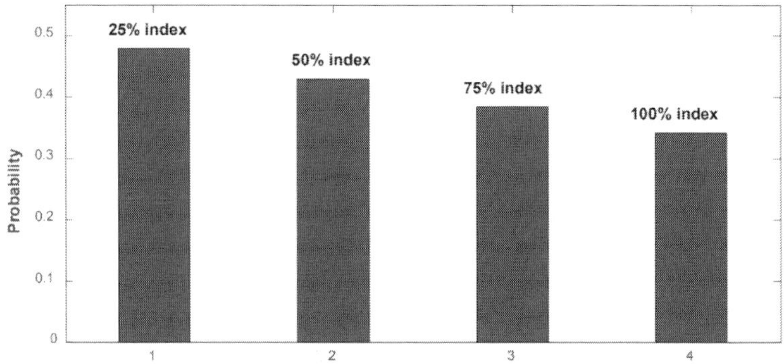

Figure 2. The probability indices of solar energy alone.

if the number of PV panels is increased, the percentage of the load that can be supplied by the solar will definitely increases. The 50% index value is 0.43 which is lower than 25% index. This means that the solar energy can supply 50% of the energy demand has a percentage of 0.43 during the month. Figure 2 shows that solar energy can supply the whole load during 0.34 of the month i.e., during 10 days.

Wind Energy

Since wind energy is of intermittent and fluctuating nature compared with the solar energy, therefore, it is hard to predict and anticipate.

Figure 3 shows the wind energy produced from wind turbines alone (no solar PV panels). It shows the highly intermittent nature of wind energy. If we compare Figure 3 with Figure 1, we can see that solar energy is more stable and consistent. From Figure 3, the wind energy can supply the whole load for only short periods between the $7^{th}$and 10th days. In most of the month time the wind energy is lower than the energy demand.

Figure 4 shows the values of the proposed indices for wind energy alone. It shows that the probability that wind energy will supply 25% of the demand during the month is 0.16. Therefore, we can rely on wind power (with 2000 wind turbines) to supply 25% of the load approximately 16% of the month (4.8 days) which is very small compared to that of solar energy. Therefore, it can be concluded that wind energy in this case is less reliable than solar energy. The values of 50% index, 75% index and 100% index is 0.065, 0.031 and 0.015 respectively.

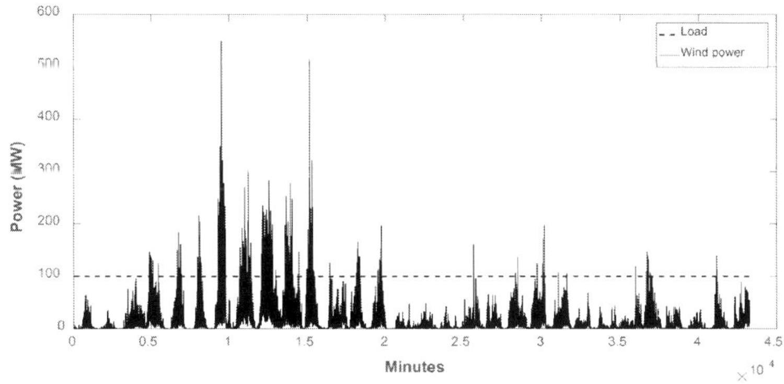

Figure 3. The wind energy produced during the month minutes.

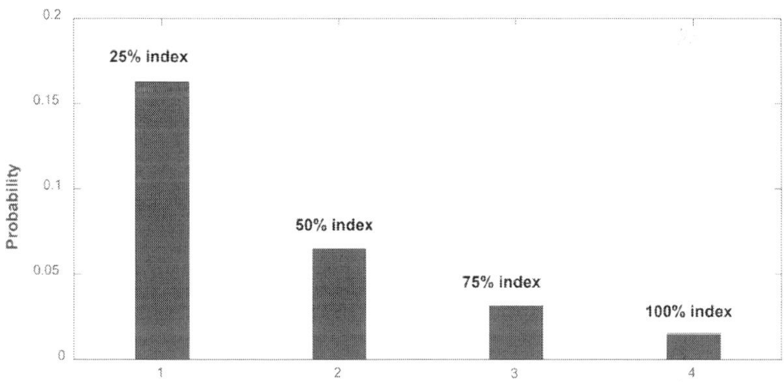

Figure 4. The probability indices of wind energy alone.

Hybrid System (Solar-Wind Energy)

Figure 5 shows the case of hybrid generation. Both wind and solar are utilized to supply the load. It can be seen that the wind energy fills some gaps left by the solar energy in the area under the load line. This means more load will be supplied and the reliability will enhances.

Figure 6 shows the reliability (availability) indices values of the hybrid system. The value of the 25% index is 0.55 which is larger than that of solar and wind alone. Therefore, 25% of the demand energy can be supplied by this renewable system during 0.55 of the month (16.5 days). The values of the 50%, 75% and 100% indices are 0.46, 0.41 0.37 respectively. That means this systems can supply 50%, 75% and 100% of the load during 0.46, 0.41 and 0.37 of the month (13.8, 12.3 and 11.1 days).

It can be concluded that hybridizing different renewable resources improves the system reliability and energy availability.

Figure 7 shows comparison between the three cases, where S denotes the solar energy, W denotes the wind energy and H the hybrid energy. It highlights the fact that reliability of renewable power systems can be improved by using more than one type of energy resources.

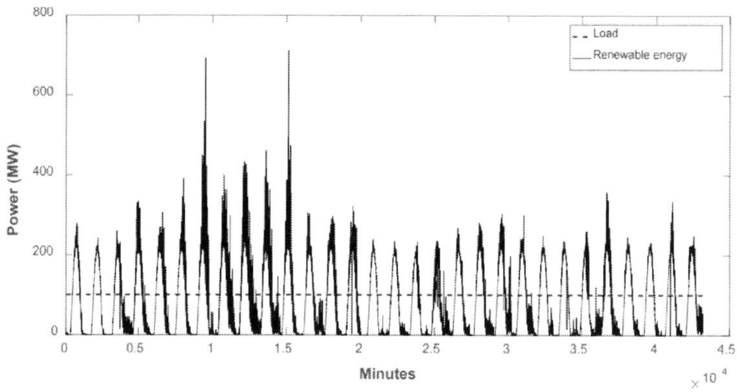

Figure 5. The renewable hybrid energy produced during the month minutes.

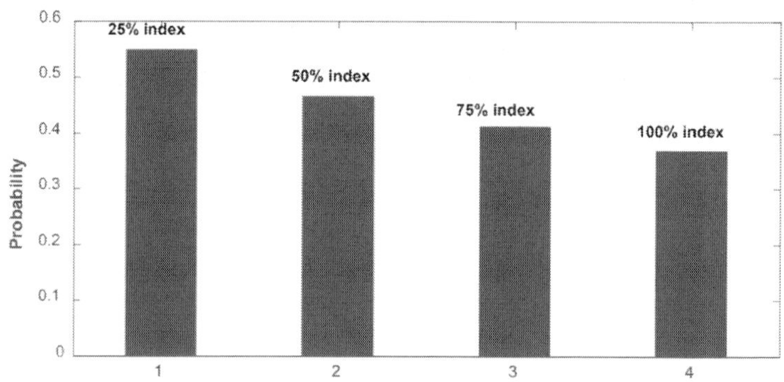

Figure 6. The probability indices of the hybrid system.

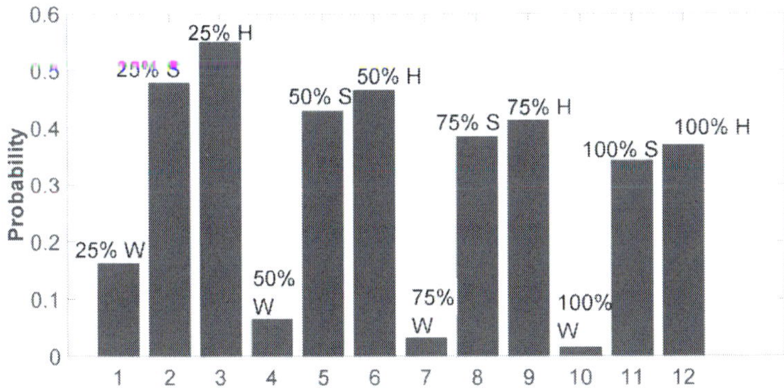

Figure 7. Comparison the three cases, solar alone, wind alone and hybrid systems.

CONCLUSION

This work presented a simulation method based on a statistical method to evaluate the availability of renewable energy power that can be quantified in supplying the load in advance over a period reliably and adequately. It was shown that, hybridizing different renewable resources (i.e. wind and solar) will improve the system's reliability and dependability. The benefit of the approach adopted and used in this work is that it can utilize historical data rather than forecasted ones.

REFERENCES

1. Billinton, R. and Karki, R. (2001) Maintaining Supply Reliability of Small Isolated Power Systems Using Renewable Energy. IEE Proceedings-Generation, Transmission and Distribution, 148, 530-534. https://doi.org/10.1049/ip-gtd:20010562
2. Stember, L., Huss, W. and Bridgman, M. (1982) A Methodology for Photovoltaic System Reliability & Economic Analysis. IEEE Transactions on Reliability, 31, 296-303. https://doi.org/10.1109/TR.1982.5221344
3. Longrigg, P. (1978) System Design and Economic Analysis of a Solar Photovoltiac Power Supply. International Telephone Energy Conference, INTELEC'78, Washington DC, 114-120.
4. Dhople, S.V. and Dominguez-Garcia, A.D. (2012) Estimation of Photovoltaic System Reliability and Performance Metrics. IEEE Transactions on Power Systems, 27, 554-563. https://doi.org/10.1109/TPWRS.2011.2165088
5. Shimura, S., Herrero, R., Zuffo, M.K. and Grimoni, J.A.B. (2016) Production Costs Estimation in Photovoltaic Power Plants Using Reliability. Solar Energy, 133, 294-304. https://doi.org/10.1016/j.solener.2016.03.070
6. Hosseini, S.J.A.-D., Moradian, M., Shahinzadeh, H. and Ahmadi, S. (2018) Optimal Placement of Distributed Generators with Regard to Reliability Assessment Using Virus Colony Search Algorithm. International Journal of Renewable Energy Research (IJRER), 8, 714-723.
7. Abouzahr, I. and Ramakumar, R. (1991) Loss of Power Supply Probability of Stand-Alone Photovoltaic Systems: A Closed Form Solution Approach. IEEE Transactions on Energy Conversion, 6, 1-11. https://doi.org/10.1109/60.73783
8. Maghraby, H., Shwehdi, M. and Al-Bassam, G.K. (2002) Probabilistic Assessment of photovoltaic (PV) Generation Systems. IEEE Transactions on Power Systems, 17, 205-208. https://doi.org/10.1109/59.982215
9. Paliwal, P., Patidar, N. and Nema, R. (2014) A Novel Method for Reliability Assessment of Autonomous PV-Wind-Storage System Using Probabilistic Storage Model. International Journal of Electrical

Power & Energy Systems, 55, 692-703. https://doi.org/10.1016/j.ijepes.2013.10.010

10. Nikmehr, N. and Ravadanegh, S.N. (2016) Reliability Evaluation of Multi-Microgrids Considering Optimal Operation of Small Scale Energy Zones under Load-Generation Uncertainties. International Journal of Electrical Power & Energy Systems, 78, 80-87. https://doi.org/10.1016/j.ijepes.2015.11.094
11. Wagner, F. (2014) Renewable in Future Power Systems. Springer, Essen. https://doi.org/10.1007/978-3-319-05780-4
12. Jarass, L., Obermair, G.M. and Voigt, W. (2009) Windenergie: Zuverlässige Integration in die Energieversorgung. Springer-Verlag, Berlin. https://doi.org/10.1007/978-3-540-85253-7

11

A New Approach for Converting Renewable Energy to Stable Energy

Mohamed Talaat[1], Reda Edris[2], Naglaa Ibrahim[2], Fatma Omar[2], Mohamed Ibrahim[2]

[1]Electrical Power and Machines, Faculty of Engineering, Zagazig University, Zagazig, Egypt
[2]Electrical and Computer Engineering, Higher Technological Institute, 10th of Ramadan City, Egypt

ABSTRACT

A renewable energy plant which relies on wind speed or solar insolation is unreliable because of the stochastic nature of weather patterns. It is theorized that by using multiple renewable energy plants in separate areas of a region, the different weather conditions might approach a probabilistically independent relationship. The goal of this paper is to utilize the power system technology to help disseminate wind and solar power systems to get a stable energy. A new approach to get appropriate stable energy is

Citation: M. Talaat, R. Edris, N. Ibrahim, F. Omar and M. Ibrahim, "A New Approach for Converting Renewable Energy to Stable Energy," Engineering, Vol. 5 No. 10A, 2013, pp. 27-33. doi: 10.4236/eng.2013.510A005.

achieved by using the interrupted energy that obtained from wind farm and solar insolation. This is achieved by lifting water to a higher level with appropriate pumps and storing it in the form of potential energy. Then a stable energy is obtained by reliving water to the lower level. In this paper, the efficiency obtained from the renewable energy is compared with that obtained from traditional ones. An experimental model to simulate the process of converting the renewable energy to a stable energy is presented. The obtained results from experimental model explained that the renewable energy can be converted to a stable one with high efficiency.

Keywords: Renewable Energy; Stable Energy; Energy Stored; Wind Farm; Solar Energy; Application of Renewable Energy

INTRODUCTION

Renewable energy sources, such as wind and solar, have vast potential to reduce dependence on fossil fuels and greenhouse gas emissions in the electric sector. Climate change concerns, state initiatives including renewable portfolio standards, and consumer efforts are resulting in increased deployments of both technologies. Both solar PV and wind energy have variable and uncertain sometimes referred to as "intermittent" output, which are unlike the dispatchable sources used for the majority of electricity generation [1-2].

The variability of these sources has led to concerns regarding the reliability of an electric grid that derives a large fraction of its energy from these sources as well as the cost of reliably integrating large amounts of variable generation into the electric grid [3-5]. Because the wind doesn't always blow and the sun doesn't always shine at any given location, there has been an increased call for the deployment of energy storage as an essential component of future energy systems that use large amounts of variable renewable resources. However, this often-characterized "need" for energy storage to enable renewable integration is actually an economic question [2,4].

To determine the potential role of storage in the grid of the future, it is important to examine the technical and economic impacts of variable renewable energy sources. It is also important to examine the economics of a variety of potentially competing technologies including demand response, transmission, flexible generation, and improved operational practices. While there are clear benefits of using energy storage to enable greater penetration of wind and solar, it is important to consider the potential role of energy storage in relation to the needs of the electric power system as a whole [2,5].

In this paper, the role of energy storage in the electricity grid has been explored, focusing on the effects of variable renewable sources (primarily wind and solar energy), a new technique of simulation like FEM and CSM are used to simulate the field and energy [6-7]. The goal is to utilize and forecast the power system technology [8] to help disseminate wind and solar power systems to get a stable energy. This is achieved by lifting water and storing electrical energy in the image of potential energy to lift water to a higher level with appropriate pumps then obtain a stable energy by releasing electricity impulsively water to the lower level. The current role that energy storage plays in meeting the varying electricity demand is important. The impact of variable renewable on the grid is then discussed, including how these energy sources will require a variety of enabling techniques and technologies to reach their full potential. Finally, the potential role of several forms of enabling technologies including energy storage has been evaluated.

ENERGY STORAGE TECHNOLOGIES

A variety of technologies are available for storage of energy in the power system [9,10]. When identifying the most relevant storage solutions it is necessary to include considerations on many relevant parameters, such as: cost, lifetime, reliability, size, storage capacity and environmental impact. All these parameters should be evaluated against the potential benefits of adding storage in order to reach a decision on which type of storage should be added. There may also be cases where the value of adding storage is not large enough to justify such an investment.

Energy storage technologies for power applications can be divided according to the form of energy stored: Mechanical, electro-chemical, electromagnetic, or thermal storage.

Mechanical storage includes pumped hydro storage, CAES and flywheels. Electrochemical storage includes all types of batteries and fuel cells, and electromagnetic storage includes super capacitors and superconducting magnetic energy storage. Each technology has certain attributes with regard to for example storage capacity, power, reaction time and cost [10].

Figure 1 shows some of the most relevant storage technologies, grouped according to the form of stored energy as well as energy storage capacity. The medium capacity storage technologies seem very relevant for storage in relation to wind power. The medium capacity storage technologies are primarily batteries and flow batteries, which all have the advantage in relation to wind power plants that they are modular and scalable.

High-Energy Batteries

For many batteries, there is considerable overlap between energy management and the shorter-term applications discussed previously. Furthermore, batteries can generally provide rapid response, which means that batteries "designed" for energy management can potentially provide services over all the applications and timescales discussed.

Several battery technologies have been demonstrated or deployed for energy management applications. In addition to the chemistries discussed previously, the commercially available batteries targeted to energy management include two general types: high-temperature batteries and liquid electrolyte flow batteries.

Pumped Hydro Storage (PHS)

Pumped hydro is the only energy storage technology deployed on a gigawatt scale in the United States and worldwide. Many of the sites store 10 hours or more, make the technology useful for load leveling. PHS is also used for ancillary services. PHS uses conventional pumps and turbines and requires a significant amount of land and water for the upper and lower reservoirs. PHS plants can achieve round-trip efficiencies that exceed 75% and may have capacities that exceed 20 hours of discharge capacity. Environmental regulations may limit large-scale above-ground PHS development. However, given the high round-trip efficiencies, proven technology, and low cost compared to most alternatives, conventional PHS is still being pursued in a number of locations.

Compressed Air Energy Storage (CAES)

CAES technology is based on conventional gas turbine technology and uses the elastic potential energy of compressed air. Energy is stored by compressing air in an airtight underground storage cavern. To extract the stored energy, compressed air is drawn from the storage vessel, heated, and then expanded through a high-pressure turbine that captures some of the energy in the compressed air. The air is then mixed with fuel and combusted, with the exhaust expanded through a low-pressure gas turbine. The turbines are connected to an electrical generator.

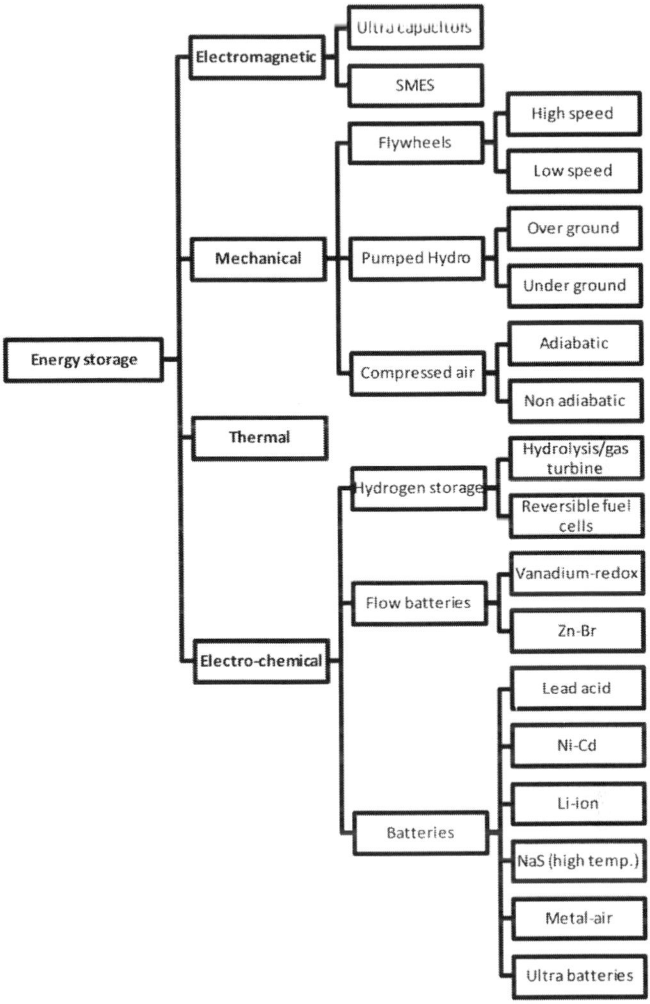

Figure 1. Energy storage technologies grouped according to form of energy as well as energy storage capacity. Typical timescales have been indicated [10].

The primary disadvantages of CAES are the need for an underground cavern and its reliance on fossil fuels. Alternative configurations for CAES have been proposed using manufactured above-ground vessels, new turbine designs to reduce fossil fuel use, or designs that re-use the heat of compression and avoid fuel use altogether.

Thermal Energy Storage

Thermal energy storage is sometimes ignored as an electricity storage technology because it typically is not used to store and then discharge electricity directly. However, in some applications, thermal storage can be functionally equivalent to electricity storage. One example is storing thermal energy from the sun that is later converted into electricity in a conventional thermal generator. Another example is converting electricity into a form of thermal energy that later substitutes for electricity use such as electric cooling or heating.

In the low capacity end, ultra capacitors may be of relevance in relation to wind power conditioning. Hydrogen fuel cells may also be relevant, both as mediumand high capacity storage.

Due to short-term energy market closure delay, dispatch levels of wind farm-energy storage system should be decided according to wind power LF [8]. The forecast accuracy and dispatch strategy are both of importance to energy storage sizing. Next, wind power forecast method and the dispatch strategy for minimizing size of energy storage system will be studied.

The low capacity storage technologies seem less relevant in relation to overall improvement of wind energy quality because of high cost pr. unit stored energy and relatively short storage time scale. The very high capacity technologies, PH storage and CAES, involve large investments and civil engineering efforts, as well as special requirements with regard to placement. But these technologies may well be the best solution in relation to large scale storage, where a few major energy storage facilities ensure overall power system stability. Especially in cases where hydro-electric plants with unused storage capacity already exist, large benefits may be obtained by combining them with wind power plants.

MATHEMATICAL MODEL

The PM DC motor is the one that is most commonly used in PVPS. It does not need field current and the magnetic field is provided separately by the permanent magnet. When a voltage V_t is applied at time t across the armature winding including an inductance L_a and resistance R_a. Then the relationship between V_b L_a and R_a can be expressed as.

$$V_t(t) = E_b(t) + R_a I_a(t) + L_a \frac{dI_a(t)}{dt} \quad (1)$$

where; I_a is the motor current and E_b is the internally induced voltage by the motor. Which is proportional to motor rotational speed ω (red/sec) such that,

$$E_b(t) = K_E \omega(t) \quad (2)$$

where; K_E is the voltage constant in V.sec/red. Combining Equations (1) and (2) gives

$$V_t(t) = K_E \omega(t) + R_a I_a(t) + L_a \frac{dI_a(t)}{dt} \quad (3)$$

The motor develops an electric torque T_d which basically depends on the motor current

$$T_d(t) = K_t I_a(t) \quad (4)$$

where; K_t is the torque constant (N·m/A).

The lead torque is the pump torque T_p which depends on the pump type, as will be seen later .If the motor pump coupling losses are neglected, the dynamic equation of the motor and load is,

$$T_d(t) = T_p + J \frac{d\omega(t)}{dt} \quad (5)$$

where; J is the moment of inertia of motor-pump system (N·m·sec^2/red). Combining Equations (4) and (5) gives,

$$K_t I_a(t) = T_p + J \frac{d\omega(t)}{dt} \quad (6)$$

The two first order differential Equations (3) and (6) represent the dynamic model of the system. The dynamic performance can be obtained by solving these equations numerically to obtain the instantaneous values of the motor current and speed.

Under steady sate conditions Equations (3) and (6) become.

$$V_t(t) = K_E \omega(t) + R_a I_a \quad (7)$$

$$K_t I_a(t) - T_p \quad (8)$$

Equations (7) and (8) represent the mathematical model of the system under steady conditions. The solution of these equations at any climatic conditions gives the operating parameters of the motor current and speed.

$$V_t = K_E \omega + R_a \frac{T_p}{K_t} \tag{9}$$

The torque constant

$$k = \frac{V_t}{\text{speed}} \tag{10}$$

Hence the torque is $T_d(t) = kI_a(t)$ and the output power is given as

$$P_{\text{out}} = T_d(t) \times \frac{2\pi \times \text{speed}}{60} \tag{11}$$

EXPERIMENTAL MODEL

An experimental model was established to simulate the mechanism of utilize the energy from wind farm and solar energy and convert this energy to a stable energy using a water pump see **Figure 2**.

A model of wind turbine or photovoltaic cell was connected to a DC pump through a PLC device. The DC pump was 24 V and 400 mA. The pump received the interrupted energy from wind if it was active or PV cell if it was active, the activation decided through the PLC sensor device. The water pump using these interrupted energy to left the water to a high tank, which use this water as a hydraulic high dam to convert this energy to a stable energy.

The performance of the wind turbine observation depends on the amount of emf produced with variable speed of wind. Also the characteristics of the DC pump depend on the amount of current taken from pump at each speed to lift the water to high tank.

A new microcontroller circuit is designed for measuring the values of current, voltage, and speeds, see Figure 3.

RESULTS

The torque constant according to Equation (10) is given as

$$k = \frac{17}{2105 * \frac{2\pi}{60}} = 0.0771 \frac{\text{Volts}}{\text{rad/sec}} = 0.0771 \frac{\text{N} \cdot \text{m}}{\text{A}}.$$

where, V_t = emf – armature drop = 24 – 7 = 17 Volts, and the corresponding angular speed is given as 2105 rpm, as given from Table 1 at no load of pump.

The voltage output from wind turbine, and the current delivered to the pump at different speed of wind are given in Table 2. Also the torque calculated from Equation (4), and the calculated powers from Equation (11) are calculated in the same table.

Figure 4 shows the variation of output voltage from wind turbine with wind speed at no load. Also Figure 5 shows the variation of delivered current to the DC pump with each wind turbine speed.

The variation of output torque with output current of wind turbine is shown in Figure 6. Also Figure 7 shows the variation of DC pump current with wind turbine speed. Figure 8” target=”_self”> Figure 8 gives the variation of wind efficiency with wind speed.

Finally, the overall efficiency of the wind and pump is calculated and shown in Figure 9, which gives a stable power.

DISCUSSION

Tables 1 and 2, also Figures 4-7 show that the energy obtained from the wind turbine with no load or to the DC pump is not stable and interrupted, so a curve fitting is used to get a stable curve.

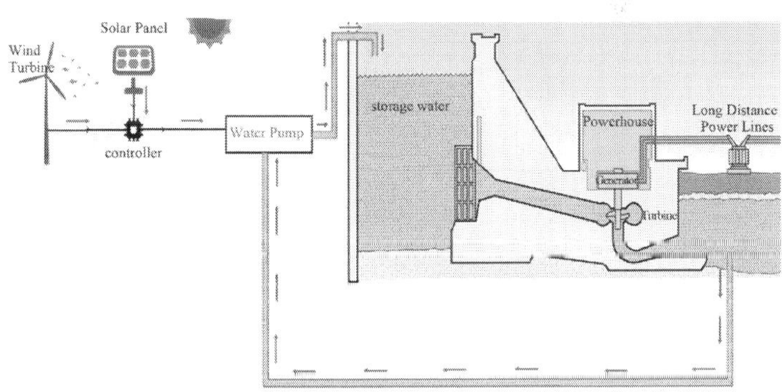

Figure 2. Energy storage simulation from wind and solar.

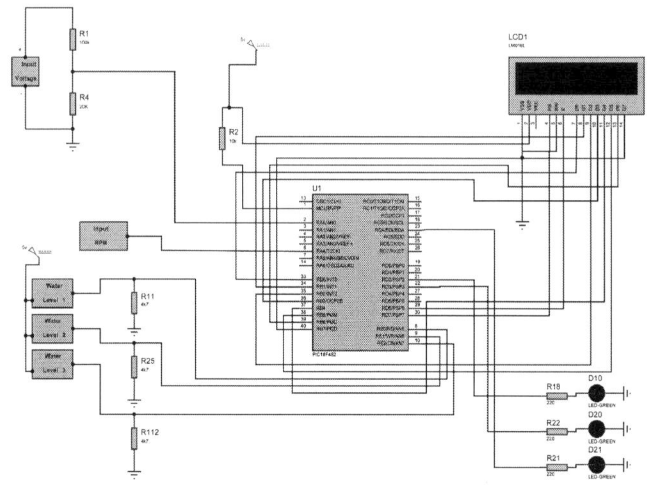

Figure 3. Microcontroller circuit used for measuring voltage, current, and speed.

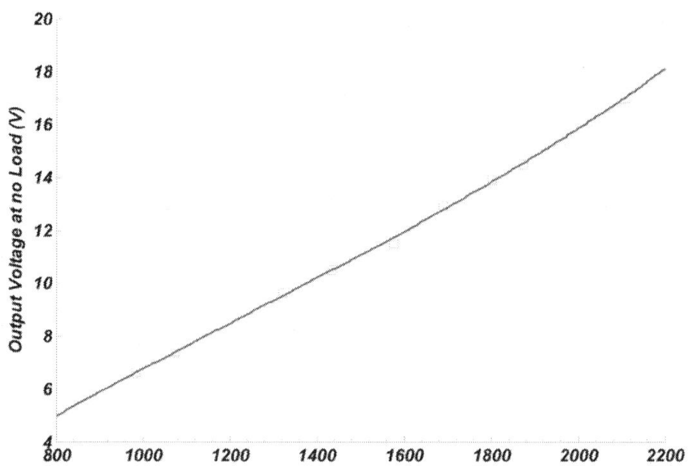

Figure 4. Variation of output no load voltage with wind turbine speed.

Also **Figure 8** explained that the variation of efficiency of wind with wind speed is not stable. So instead to use this unstable energy for power generation which will be unstable also, the new technique uses this unstable energy for lifting the water to high tank to generate a stable energy, according to Figure 9 the overall efficiency gives a stable energy.

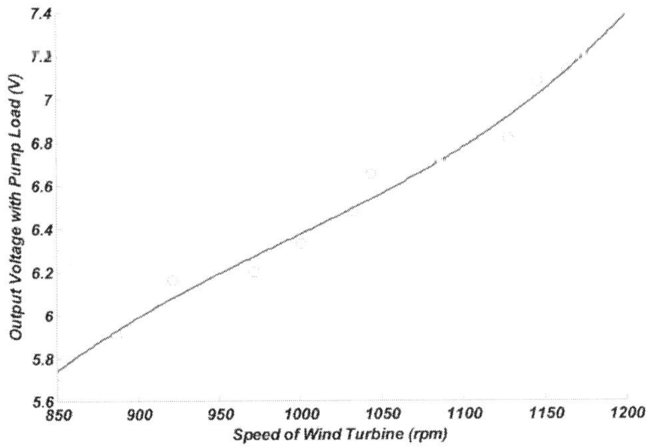

Figure 5. Variation of output voltage with wind turbine speed.

CONCLUSION

An experimental model to simulate the process of converting the renewable energy to stable energy is presented. A new approach to get appropriate stable energy is achieved by using the interrupted energy that obtained from wind farm and solar insolation. This is achieved by

Table 1. Variation of output voltage with wind turbine speed at no load.

Wind Turbine Speed (rpm)	Output Voltage (V)
985	6.6
1076	7.4
1323	9.6
1435	10.5
1577	11.5
1690	12.9
1800	13.9
1874	14.5
2105	17

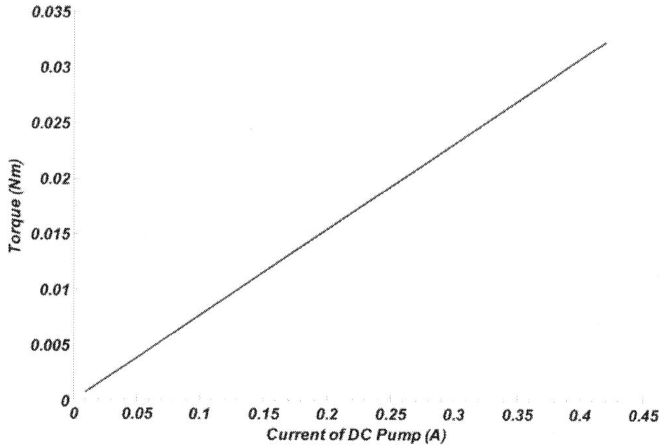

Figure 6. Variation of output torque with output current.

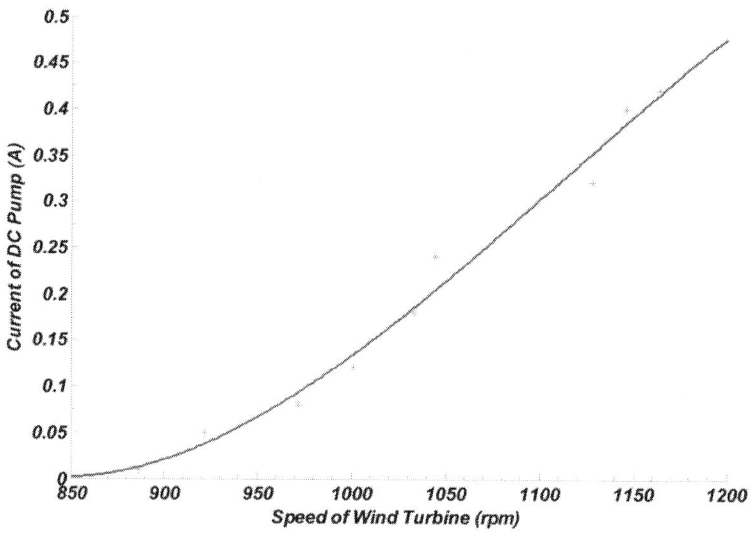

Figure 7. Variation of DC pump current with wind turbine speed.

lifting water and store electrical energy in the image of potential energy to lift water to a higher level with appropriate pumps then obtain a stable energy by releasing

Table 2. Variation of delivered current to DC pump with wind turbine speed.

Wind Turbine Speed (rpm)	Output Voltage (V)	Current (A)	Output Power (W)	Torque (Nm)
887	5.9	0.01	0.059	0.0008
922	6.16	0.05	0.308	0.0039
972	6.20	0.08	0.496	0.0062
1001	6.33	0.12	0.7596	0.0093
1033	6.48	0.18	1.1664	0.0139
1044	6.65	0.24	1.596	0.0185
1128	6.82	0.32	2.1824	0.0247
1146	7.08	0.40	2.832	0.0308
1164	7.13	0.42	2.9946	0.0324

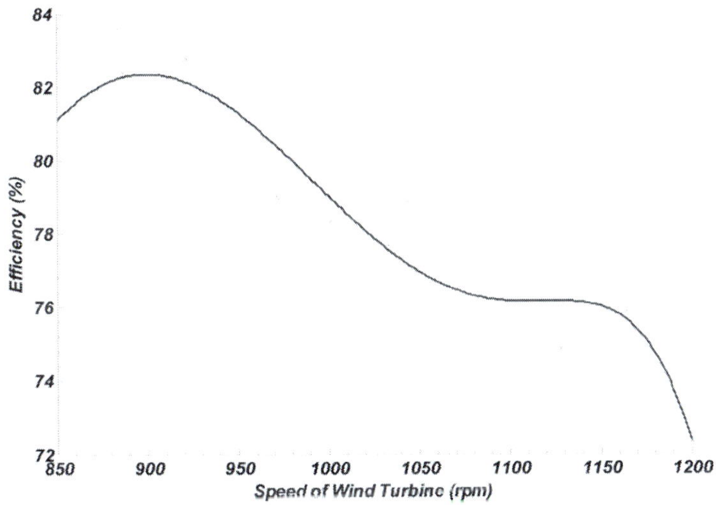

Figure 8. Variation of wind efficiency with wind turbine speed.

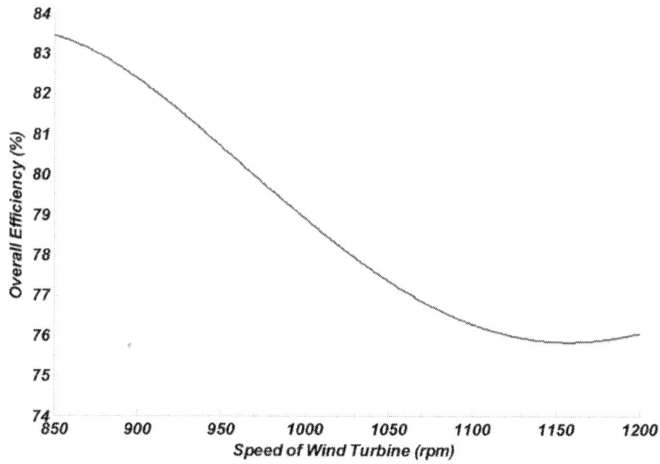

Figure 9. Variation of overall efficiency with wind turbine speed.

Mohamed Talaat, Reda Edris, Naglaa Ibrahim, Fatma Omar, Mohamed Ibrahim electricity impulsively water to the lower level. The obtained results from experimental model explained that the renewable energy can be converted to a stable one with high efficiency.

Nomenclature A

PV: Photovoltaic FEM: Finite Element Methods CSM: Charge Simulation Method LF: Load Forecasting CAES: Compressed Air Energy Storage PH: Pumped Hydro PM: Permanent Magnet PHS: Pumped Hydro Storage PVPS: Photovoltaic Pump Storage

REFERENCES

1. World Wind Energy Association, "Highlights of the World Wind Energy Report," 2009. http://www.wwindea.org/home/index.php
2. "The Role of Energy Storage with Renewable Electricity Generation," Technical Report NREL/TP-6A2-47187, 2010. http://www.osti.gov/bridge
3. L. L. Freris, "Wind Energy Conversion Systems," Prentice Hall, Upper Saddle River, 1990.
4. B. C. Ummels, E. Pelgrum and W. L. Kling: "Integration of Large-Scale Wind Power and Use of Energy Storage in the Netherlands' Electricity Supply," IET Renewable Power Generation, Vol. 2. No. 1, 2008, pp. 34-46.
5. E. Spahic, G. Balzer, B. Hellmich and W. Münch, "Wind Energy Storages—Possibilities," IEEE PowerTech, 2007.
6. M. Talaat and A. El-Zein, "A Numerical Model of Streamlines in Coplanar Electrodes Induced by Non-Uniform Electric Field," Journal of Electrostatics, Vol. 71, No. 3, 2013, pp. 312-318. http://dx.doi.org/10.1016/j.elstat.2012.12.034
7. M. Talaat, "Charge Simulation Modeling for Calculation of Electrically Induced Human Body Currents," IEEE Annual Report Conference on Electrical Insulation and Dielectric Phenomena CEIDP, West Lafayette, 17-20 October 2010, pp. 644-647.
8. M. A. Farahat and M. Talaat, "The Using of Curve Fitting Prediction Optimized by Genetic Algorithms for Short- Term Load Forecasting," International Review of Electrical Engineering (IREE), Vol. 7, No. 6, 2012, pp. 6209- 6215.
9. M. Swierczynsky, R. Teodorescu, C. N. Rasmussen, P. Rodriguez and H. Vikelgaard, "Storage Possibilities for Enabling Higher Wind Energy Penetration," EPE Wind Energy Chapter Symposium, Stafford, 15-16 April 2010.
10. C. N. Rasmussen, "Energy Storage for Improvement of Wind Power Characteristics," IEEE PowerTech, Trondheim, 2011.

12

Powering Renewable Programs: The Utility Perspective

Nicole Griffin, Athens Gomes Silaban

UtiliWorks Consulting, Baton Rouge, LA, USA.

ABSTRACT

In order to make renewable energy projects successful, there are many factors that utilities need to consider. These include policy drivers, assessing what renewable technologies it will employ, identifying the rates and pricing incentives that could be made available, and how customers can be better engaged. Utilities have created renewable programs with varying degrees of customer participation: some have taken the initiative to provide customers with 100% renewable generated power, others rely exclusively on customers to participate to meet renewable energy goals and the last alternative is a blend of both in which utilities offer customers the option to purchase renewable power matches or install and generate their own renewable

Citation: Griffin, N. and Silaban, A. (2016), "Powering Renewable Programs: The Utility Perspective". Open Journal of Energy Efficiency, 5, 148-159. doi: 10.4236/ojee.2016.54013.

power. Overall, the economics of solar and wind technologies are trending in the upward direction—the costs of the technologies are decreasing and the yields are getting higher. Better still, predictive modeling, energy storage and the plethora of research in this area will only make the prospects of integrating renewables more viable.

Keywords: Renewable Energy, Electric Utility, Solar, Wind, Energy Storage

INTRODUCTION

When considering how to build out a renewable energy portfolio, there are a lot of factors and many points of view to consider. For purposes of this narrative, we will take the point of view of the electric utility itself.

The interest in renewable energy is rapidly increasing with many countries proposing ambitious roadmaps of their power sector renewable integration. So how does an electric utility begin building its renewable roadmap? First, the utility needs to identify what types of renewable technologies are suited for its environment. Using the US, for this example, we of course know that there are some regions that are well suited for hydro-electricity, such as the infamous Hoover Dam area. In drier areas of the Midwest where there are not steady tides or waves that can be used to produce electricity, hydroelectric power is simply not an option. Solar and wind represent a different scenario, whereby all areas have varying degrees of either. The questions for solar and wind energy then become:

- How much energy can we generate?
- Is it profitable to utilize these technologies?
- Is it policy driven?

DRIVERS FOR SOLAR AND WIND GENERATION

Let's start on the concept of how much renewable energy we can generate. Firstly, wind power. Of the approximately 48,800 operating utility-scale wind turbines, majority of them correlate to where the capacity is on the map presented in Figure 1 [1] . Traditionally in the US, the Midwest has the largest wind capacity—as farmlands in plains states can produce large amount of winds [2] . In terms of actual production, the state of Texas far surpasses wind generation of any state at 17,711 MW, with Indiana coming in second at 6209 MW and California third at 5662 MW.

In the United States alone, the additions of new wind turbines and infrastructure increased the power capacity for the year 2015 by 41% and powered 88,000 jobs [3] . The DOE's "2014 Wind Technologies Market Report" shows that wind turbine prices have fallen drastically, 20% - 40% lower than their highs in 2008. Larger rotors are advancing in technology with increasing capacity and efficiencies. Concurrently, wind energy is selling at lower and lower prices, hitting 2.5 cents per kWh per the report as compared to the national average 11 cents per kWh [4] .

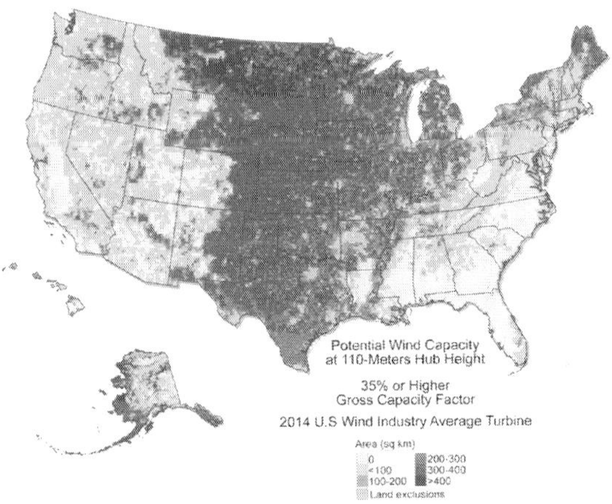

Figure 1. US map of potential wind capacity, by state (2014).

As impressive as the investments in wind energy have been, the global investments in solar have been even greater in recent years (Figure 2).

One of the reasons that solar is gaining so much popularity is due to its versatility. As opposed to wind power, solar can be easily implemented at local business and residential premises. The graphic below shows how along with utility-owned photovoltaics (PV), homes and businesses have a high adoption rate. One of the largest drivers is the continued decrease in costs of solar panels, at about an average of 10% each year since the 1980s [6] . Along with a decrease in costs, the panels are becoming more efficient and more aesthetically pleasing (thinner and blend in better with rooftops). The amount of solar power installed in the US has increased with this trend at more than 23 times from 1.2 GW in 2008 to an estimated 27.4 GW at the end of 2015 [7] . The growing industry also has had an extremely positive effect on the US workforce. The Solar Foundation's National Solar Jobs Census 2015, revealed "that in 2015 the solar workforce grew at a rate 12 times faster than the overall economy and now boasts 208,859 solar workers [8] ."

As presented in Figure 3, the estimated PV installation is estimated to decrease in 2017, while residential and non-residential PV are projected to increase; the utility PV installation is anticipated to experience decrease due to the Solar Investment Tax Credit (ITC) project completion deadline extension [10] . Installations are expected to pick up again in 2019.

As valuable as solar and wind power may prove to be, they are naturally unreliable— the wind doesn't always blow, and the sun doesn't always shine. The energy that can be captured is intermittent. The conundrum, especially in the instance of residential solar power for example is that when power is being produced (mid) is often opposite of when consumers need it (after work hours in the evening). This defies the premise that electrical supply and demand must be equal at any given moment.

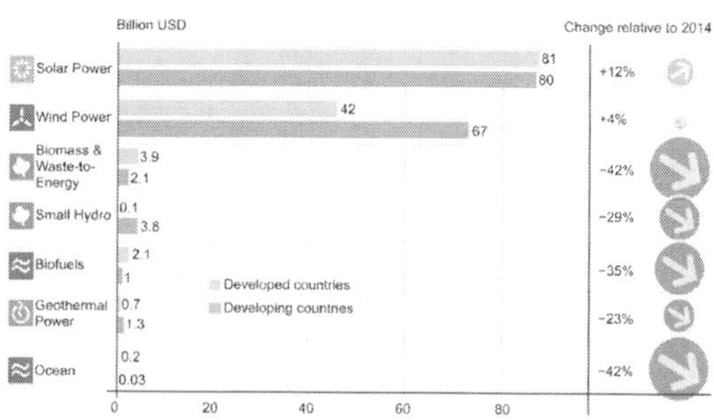

Figure 2. New investments in renewable energy, Developed & developing countries (2015) [5].

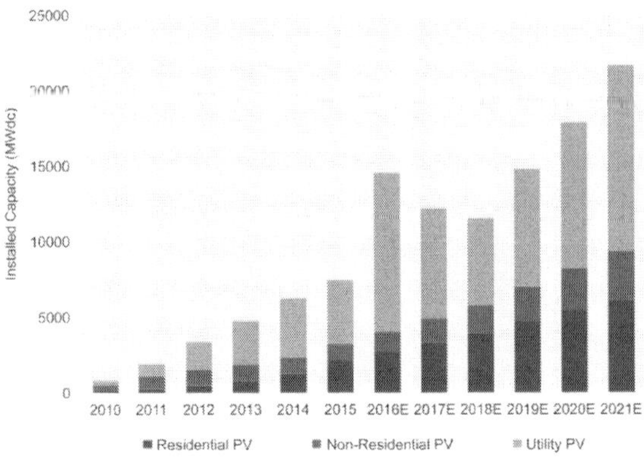

Figure 3. Installed photovoltaic & projected photovoltaic installation in the US [9] .

Thus, what is required to make these renewables truly effective is energy storage. Energy storage helps balances the act of supply and demand by storing excess energy and distributing that energy when needed, creating a more flexible and reliable grid system. Deploying energy storage will also reduce and potentially eliminate the need to generate or buy power during high (peak) demand times. Conventional energy storage methods utilized by utilities include pumped hydroelectric (largest storage system operating today in the US), thermal storage, compressed air and hydrogen [11] . However the most promising energy storage option may be the use of battery cells, where electricity is converted to chemical potential for storage and generated when needed. Advanced batteries are a major focus of research into improving the efficiency of battery cells, with various technologies being examined as reflected in Figure 4 [12] .

The interest in battery cells for electric utility energy storage purposes can be related to the increased interest and production of electric vehicles. The most prominent example is Tesla who started as an electric car automotive company, now they are also immersing into the energy storage sector with Tesla Powerwall (rechargeable lithium- ion battery product manufactured for home use) [13] . Energy storage technology may be immature at this point with significant costs, but they can be the answer to complementing the intermittent characteristics of renewable energy. Otherwise phrased as: How do we handle renewable generation when the demand is not there? The following graph attempts to explain how energy storage fits in the renewable energy electricity use (Figure 5).

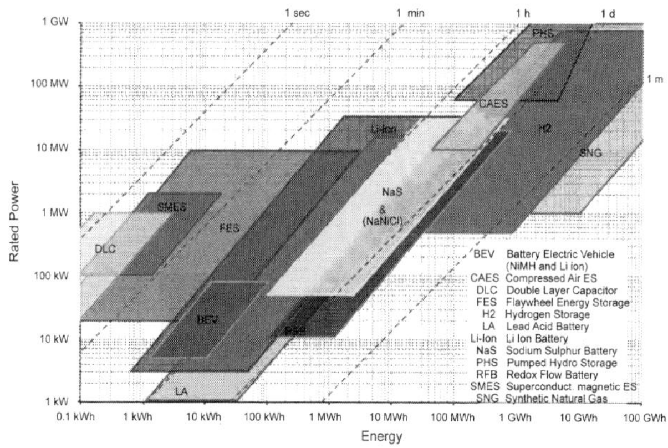

Figure 4. Characteristics of electrical energy storage technologies compared by energy, rated power & discharge time.

Predictive analytics are also an important part of the conversation especially with the variable nature of renewable energy such as solar and wind. These are remote monitoring solutions that track the performance of renewable energy productions. Predictive analytics provide a forecast of the amount of energy produced that can be directly supplied to the power grid or stored with energy storage. Predictive analytics software translates data into information which utilities can then possess to make better decisions. With predictive analytics (better yet complemented with energy storage), the utility's operational efficiency and renewable energy assets life span improves. Various predictive analytics models are developed by various vendors, from the large players of IBM, Accenture to Locus Energy, and Space-Time Insight which is represented in Figure 6 [15] .

Overall, the economics of solar and wind technologies are trending in the upward direction—the costs of the technologies are decreasing and the yields are getting higher. Better still, predictive modeling, energy storage and the plethora of research in this area will only make the prospects of integrating renewables more viable.

Renewable mandates are becoming more demanding as well, with more and more countries continuing to adopt renewable energy policies. The Renewable Energy Directive sets rules for the European Union to achieve its 20% renewables target by 2020 (Figure 7).

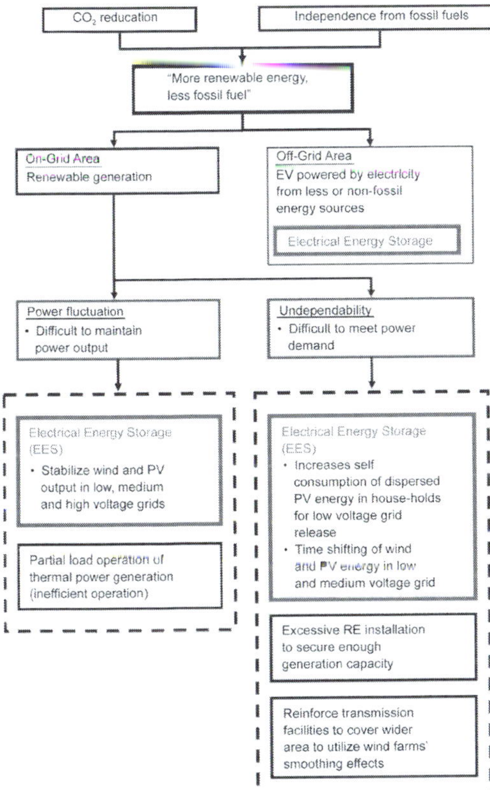

Figure 5. Presentation of electrical energy storage role in maximizing renewable energy [14] .

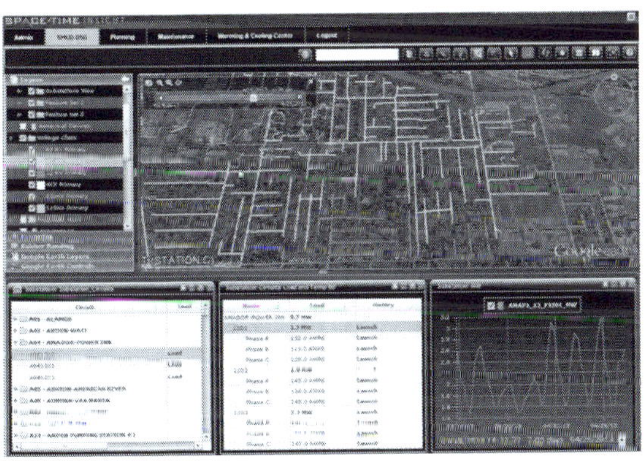

Figure 6. Screen sample of a predictive analytics software (Space-Time insight software, used for a utility's grid planning and operations).

Many of the US states are looking towards a similar 20% renewable by 2020, and 25% by 2025. California Policies tend to surmount those for the rest of the US. Recently, Governor Brown of California signed a standard into law to mandate 50% renewables by 2030, which is expected to rely heavily on solar [16] . Hawaii may have its sights set higher than California even, as the first state to set a goal of producing 100% of its electricity from renewable sources by 2045 [17] .

RENEWABLE ENERGY PROGRAM COMPARISONS

In order to meet a renewable goal of 20% or more by 2020, utilities and consumers need to be actively engaged. So the question becomes, how do we make it a reality? The utility needs to identify how this affects load curves and revenue generation. Is renewable uptake something a utility wants to do on its own or does it need to encourage customer participation; and if so to what extent? The utility also needs to determine if they will accept consumers putting power back onto the grid, and if so how much will they pay for it? Is it something they can pay at a retail or wholesale price? If the utility is seeking to increase its renewable portfolio for economic or policy reasons, are there opportunities to provide rebates?

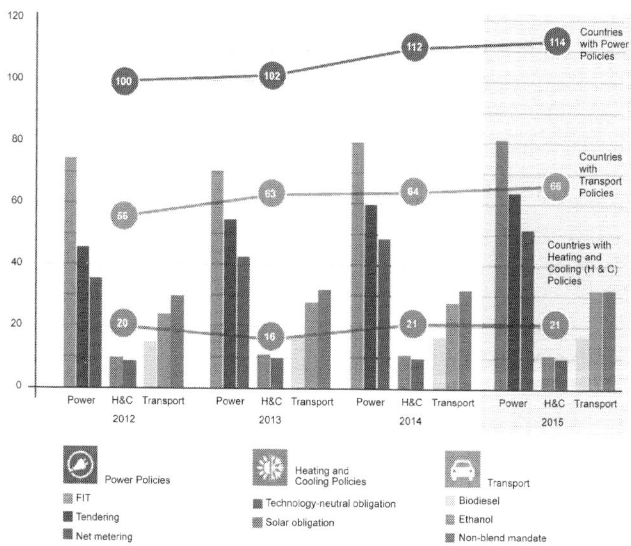

Figure 7. World's renewable energy policies overview (2012-2015).

Let's begin with a case study of a utility that has made the business decision to take control of the renewable conversation. In 2014, Georgetown, TX finalized a 150-mega- watts solar power agreement and a 144 megawatt wind power agreement that "will make the City of Georgetown one of the largest municipally-owned utilities in the US to supply its customers with 100 percent solar and wind energy [18] ." Georgetown was able to build a very positive business case by looking at cyclical generation patterns and negotiating favorable purchase power agreements. The utility is using solar and wind concurrently with the intention that solar will provide a daily afternoon supply to match peak demand period (particularly in the high demand summer times) and wind will pick up the slack off-peak hours and during the early morning. The City signed an agreement with SunEdison to provide electricity to the City through 2041 at lower overall costs than its prior wholesale agreements. The wind project comes at a 20-year contract with EDF that will provide competitive electricity prices through 2035.

What all of this means is that all customers throughout the utility's territory will be getting clean power—no residential or commercial solar or wind generation required. What's more is the utility rates for customers are expected to stay low; residential customers currently paying $20/month customer charge plus $0.0939/kWh. Georgetown utility is continuing to receive customer requests regarding rooftop solar. The decision on whether or not the utility will recognize residential or commercial wind or solar after they've achieved a 100% renewable status in 2017 is yet to be determined. Largely, customer education is expected to be a major factor in navigating through the customer owned renewable conversation.

Majority of utilities are not in a position to take renewables completely in house as Georgetown has. Another model is to offer customers the choice of generating their own renewable power or giving them the option of purchasing renewable power from their utility company. Pacific Gas & Electric (PG & E) for example, offers a "solar choice" for its customers in which they can opt to have 50% or 100% of their monthly power matched with solar power provided by the utility. Instead of purchasing and install rooftop solar, customers can pay a premium for the service through their utility provider. PG & E offers a rate calculator to assist customers in understanding the costs. For example, Figure 8 demonstrates that a customer can expect an additional cost of almost $18 a month if they opt for 100% solar choice participation at 500 kWh monthly [19] . The estimated premium is approximately $3.58 cents/kWh regardless if usage is 50% or 100%.

Like PG & E, SMUD offer its customers both the option to self-install or participate in a community solar program. SMUD's SolarShares program is setup so that customers pay a flat monthly fee to subscribe. On the SMUD website, they describe the structure further as such: the fee is based on your historical energy use and the share size you select.

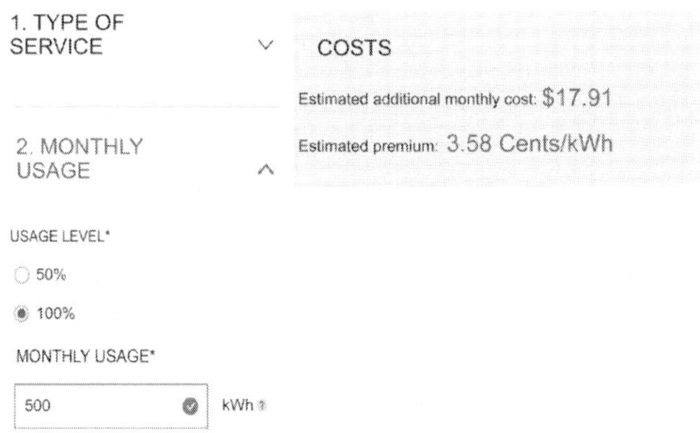

Figure 8. PG & E's Solar Rate Calculator, which quantifies expected premium costs to participate in the solar program.

You will then receive energy credits to your bill for the amount of solar power your SolarShare generates. Both the flat monthly fee and the energy credits from the solar power will be combined on your SMUD bill [20] .

Now, let's take a look at various examples of how the customer-owned renewable programs are managed. In January 2016, the California Public Utilities Commission (CPUC) voted to enact a policy that would ensure net-metered customers earn retail-rate payments for their surplus solar energy. The decision also comes with requirement for solar customers to move to time of use (TOU) rates "that charge different prices during different times of the day, to better match real-time costs of generating and transmitting energy across the grid at large [21] ."

Unlike California, many Texas utilities have not adopted net-metering policies for their end-users, nor are they required to do so. However, there are some utilities such as Texas Gas Service who offers solar water heating rebates. Other Texas utilities including Oncor, Austin Energy, CPS Energy,

El Paso, AEP TCC and AEP TNC have solar incentive options, but vary from one to the other [22] , Farmers Electric Cooperative in Iowa will pay its customers a fixed retail rate of 12.5 cents per kWh up to 100% of the accounts consumption. Above and beyond that, they are compensated 6 cents per kWh [23] .

New Jersey is working aggressively to increase solar across the state as part of its Renewable Energy Incentive Program (REIP) targeting a 30% renewable target by 2020. New Jersey is the fastest growing market for solar in the US and "one of the largest in terms of installations and installed capacity, second only to California. Much of this success is due to New Jersey's Solar Financing Model, which relies on high renewable energy standards and the use of Solar Renewable Energy Credits (SRECs) [24] ." These solar power performance payments are used to motivate New Jersey residents to produce solar power, which the utilities can virtually claim as their own to avoid paying an Alternative Compliance Payment (ACP). For every MWh the utility falls short of its renewable goal, it will pay an additional fee which for 2016 was $323 per MWh. New Jersey has a few other tactics to motivate customer uptake. These include sales tax exemption for equipment related to solar energy installation and property tax exemptions. New Jersey is also offering incentives for customer energy storage at a maximum of $300 or 30% of each project until the $3 million budget runs dry [25] . Incentives such as the ones New Jersey is offering are decreasing the payback period for residential solar. In the case of New Jersey, we can expect approximately a 6.5-year payback period [26] . Hawaiian electric has an even better payback period at 5.4-years and multiple solar options available to its customers including Customer Grid-Supply (CGS) and Customer Self-Supply (CCS), with the later not having the option to push power back onto the grid. Each of these options does include a minimum utility charge of $25 to sustain the infrastructure. Hawaii's clean energy initiative has been well received, and solar adoption more than doubled in 2014 [17] . Expectedly, one of the biggest drivers for the uptake was the utility's outreach efforts and the quality of content it has made available to its customers, such as the webpage represented in Figure 9 [27] .

Figure 9. Hawaiian Electric's Webpage, encouraging customers to implement solar panels.

The utility energy transformation is here, and renewables are undoubtedly a major component. In order to make these projects successful, there are many factors that utilities need to consider. These include policy drivers, assessing what renewable technologies it will employ, understanding how these will be integrated into the grid, identifying the rates and pricing incentives that could be made available, and how customers can be better engaged. With all factors considered, the utility will be equipped to build out a thoughtful renewable roadmap.

REFERENCES

1. American Wind Energy Association. Wind Energy Facts at a Glance. http://www.awea.org/Resources/Content.aspx?ItemNumber=5059
2. Union of Concerned Scientists. Farming the Wind: Wind Power and Agriculture. http://www.ucsusa.org/clean_energy/smart-energy-solutions/increase-renewables/farmingthe-wind-wind-power.html#.WCQv7fkrI2w
3. American Wind Energy Association. Wind Energy Facts at a Glance. http://www.awea.org/Resources/Content.aspx?ItemNumber=5059
4. Weiner, J. (2015) Study Finds That the Price of Wind Energy in the United States Is at an All-Time Low, Averaging under 2.5¢/kWh. http://newscenter.lbl.gov/2015/08/10/study-finds-that-the-price-of-wind-energy-in-the-uni ted-states-is-at-an-all-time-low-averaging-under-2-5%C2%A2kwh/
5. Ren21. (2016) Renewables 2016 Global Status Report. http://www.ren21.net/wp-content/uploads/2016/10/REN21_GSR2016_FullReport_en_11.pdf
6. Darby, M. (2016). https://www.theguardian.com/environment/2016/jan/26/solar-panel-costs-predicted-to-fall -10-a-year
7. US Department of Energy. Tackling the Hidden Costs of Rooftop Solar. http://energy.gov/science-innovation/energy-sources/renewable-energy/solar
8. US Department of Energy. (2016) Solar Energy Jobs Outpace U.S. Economy.
9. Solar Energy Industries Association (2016) Solar Market Insight Report 2016 Q2. http://www.seia.org/research-resources/solar-market-insight-report-2016-q2
10. Solar Energy Industries Association (2015) Impacts of Solar Investment Tax Credit Extension. http://www.seia.org/research-resources/impacts-solar-investment-tax-credit-extension
11. Union of Concerned Scientists. How Energy Storage Works. http://www.ucsusa.org/clean-energy/how-energy-storage-works#.V73ECEtTHIV
12. Electrical Energy Storage Project Team, International Electrotechnical Commission. Electric Energy Storage. http://www.iec.ch/whitepaper/pdf/iecWP-energystorage-LR-en.pdf
13. Tesla. Powerwall. https://www.tesla.com/powerwall

14. Kawashima, M. (2011) Overview of Electrical Power Storage. Internal Paper of Tepco.
15. Puttre, M. (2015) Analytics and Big Data Are Changing the Energy Market Map. http://solarindustrymag.com/online/issues/SI1501/FEAT_03_Analytics-And-Big-Data-AreChanging-The-Energy-Market-Map.html
16. Roselund, C. (2015) California Governor Brown Signs 50% Renewable Portfolio Standard into Law. http://www.pv-magazine.com/news/details/beitrag/california-governor-brown-signs-50-ren ewable-portfolio-standard-into-law_100021447/#ixzz4He6NKmAP
17. US Energy Information Administration. Hawaii State Profile and Energy Estimates. http://www.eia.gov/state/?sid=HI
18. Georgetown Going 100% Renewable. https://gus.georgetown.org/renewable-energy
19. PG&E's Solar Choice Rate Calculator. https://www.pge.com/en_US/residential/solar-and-vehicles/options/solar/solar-choice/ratecalculator.page
20. SolarShares®: Solar for Everyone! https://www.smud.org/en/residential/environment/solarshares.htm
21. John, J. (2016) Breaking: California's NEM 2.0 Decision Keeps Retail Rate for Rooftop Solar, Adds Time of Use. http://www.greentechmedia.com/articles/read/Californias-Net-Metering-2.0-Decision-Roof top-Solar-to-Keep-Retail-Payme
22. Blum, J. (2016) Texas Gets "F" Grade on Net-Metering Policies for Solar Power and More. http://fuelfix.com/blog/2016/01/26/texas-gets-f-grade-on-net-metering-policies-for-solar-p ower-and-more/
23. Solar Outreach Partnership. Farmer Electric Cooperative: A Small Rural Cooperative Becomes a Solar Leader. http://www.solarelectricpower.org/media/230417/SEPA031-SolarOPs-Study_FEC_0914.pdf
24. New Jersey's Clean Energy Programs. http://www.njcleanenergy.com/renewable-energy/programs/programs
25. New Jersey Renewable Electric Storage. http://www.njcleanenergy.com/renewable-energy/programs/energy-storage

26. Coulter, S. (2012). Report Finds Shorter Payback Periods on Home Solar in Ten US States http://www.prweb.com/releases/2012/3/prweb9282391.htm
27. Hawaiian Electric Company. Clean Energy Hawaii: Going Solar. https://www.hawaiianelectric.com/clean-energy-hawaii/going-solar

13

Technical Analysis and Enlightenment of Renewable Energy

Shi (Jessee) Zhang[1,2]

[1]Shenyang Agricultural University, Shenyang, China.
[2]China University of Political Science and Law, Beijing, China.

ABSTRACT

This paper analyzes China's low-carbon technology through patent data of three major renewable energy sources which include nuclear energy, wind energy, biomass energy. The data platform used in this paper is from patent retrieval and analysis system in SIPO. According to the patent application in China, this paper studies regional map of technology, domain regional technology analysis, technology trend analysis and technology life cycle in the three renewable energy. From the above analysis, this paper sums up shortcomings in development of renewable energy and policies.

Citation: Zhang, S. (2017), "Technical Analysis and Enlightenment of Renewable Energy". Low Carbon Economy,8, 106-117. doi: 10.4236/lce.2017.84009.

Keywords: China, Low Carbon technology, Renewable Energy, Patent

INTRODUCTION

Low carbon economy is one of the best models of global economic development [1] . Low carbon consumption is an important part of it. Low carbon consumption, industry, technology and urban construction are considered to be the most important content of low carbon economy. Low carbon consumption can promote the development of low-carbon industry, which makes the country into a virtuous cycle of low-carbon economy [2] . In the field of low-carbon energy technology, China's renewable resources are very rich. According to the Chinese government issued the "12th Five-Year plan" in 2011, which can see that the most important development field in low carbon economy and new energy industry are: Thermal power emission reduction, new energy vehicles, energy-saving buildings, industrial energy conservation and emission reduction, recycling of renewable resources and recycling economy, environmental protection equipment, etc. The development of China's low carbon economy is still in its infancy, and the low carbon policy system is not mature enough. The government is speeding up the establishment of supervision mechanism of energy saving and emission reduction of the voluntary agreement, perfect the supervision mechanism of energy-saving product certification, and tax relief, policy incentives and other incentives as soon as possible to establish industry standards and norms in the energy services industry, to promote the full implementation of the contract energy management [3] ! As a type of modern technology, the innovation of low carbon technology follows the general technology life cycle, to experience from basic research, technology development, demonstration projects to market promotion stage. Generally speaking, lack of core technology and R & D strength are the common characteristics and development bottlenecks of most low carbon technologies. For example: carbon dioxide capture and storage technology in the R & D demonstration phase; Wind power and solar energy technology without core technology support; Master the core technology of LED Technology [4] . According to the patent application in China, this paper studies regional map of technology, domain regional technology analysis, technology trend analysis and technology life cycle in the three renewable energy. The renewable energy technologies studied in this paper include nuclear energy, wind energy and biomass energy. The analysis data are 246 Chinese patent data in nuclear energy; The analysis data are 4134 Chinese patent data in wind energy; The analysis data are 438 Chinese patent data in Biomass energy.

LITERATURE REVIEW

China has initially formed the "renewable energy law" as the backbone of the legal system of renewable energy, in the development, utilization and management of renewable energy, the law can be realized. However, there are still insufficient laws and regulations in the legislation of renewable energy in China, and the comprehensive energy basic law has not been formulated. Some renewable energy legislation is rather deficient or the legislative force is low and the legal system is not perfect, and renewable energy laws are not closely linked to climate change. Our country should emphasize and highlight the development, utilization and management of renewable energy in the new energy basic law and the newly revised basic law of environmental protection. Combining the climate change law or the low-carbon economy promotion law with the revision of the renewable energy law, a legislative model for the implementation of a renewable energy law, the leading role of the government in promoting the development of renewable energy and the role of law in guiding and promoting the development of renewable energy. Renewable energy is the energy of sustainable development, the future energy, who masters the renewable energy, who will master the future of energy. In the face of potential energy crisis, global warming and deterioration of ecological environment, known as green energy, such as wind energy and solar energy, hydropower, biomass, geothermal energy, ocean energy, development and utilization of the renewable energy has been attention for the government and the whole society. As a mandatory and compulsory code of conduct and institutional resources, law plays an important role in promoting and protecting the sustainable development of renewable energy in China.

RESEARCH METHOD AND DATABASE

Considering many factors of China's legal development trend of new energy and environmental protection, this paper chooses three representative energy as the object of analysis the reason is: the nuclear energy is most closely related to environmental protection and energy; wind energy development is second worth of the energy in addition to solar energy; Bioenergy is the representative of energy in the micro world.

This paper analyzes the technology of China's three low-carbon renewable energy sources through China's patent data. The data platform used in this paper is patent retrieval and analysis system in SIPO. The retrieval method used in this paper is subject search, which is retrieved by searching the synonyms in the retrieval table, and the obtained data is accurate. SIPO is abbreviation of State Intellectual Property Office of the People's Republic of China, SIPO has collected the most comprehensive patent data in China. The patent retrieval and analysis system is also the most official analysis system. The author used keyword search to make the data more precise. Patent retrieval and analysis system in SIPO introduction:

Retrieval function contain regular search, table retrieval, drug thematic search, retrieval history, search results browsing, document browsing, batch download, etc. Analytical function contains quick analysis, custom analysis, advanced analysis, analysis report generation, etc.

The range of data: a collection of patent data of 103 countries, regions and organizations, as well as the citation, kin, legal status and other data, which covers the China, the United States, Japan and South Korea, Britain, France, Germany, Switzerland, Russia, the European Patent Office and the world intellectual Property Organization etc.

Data update: Chinese and foreign patent data, every Wednesday; family, legal status data, every Tuesday; citation data, updated monthly.

RESEARCH GROUND AND PURPOSE

In China, the design of renewable energy roadmap and its policy research have just started. In the development roadmap, in recent years, only the Chinese Academy of Sciences, the Chinese Academy of engineering and the Ministry of science and technology have carried out some research on the development of renewable energy technology [5] . Due to the imperfect database of China, China cannot use renewable energy technology life cycle method to comprehensively evaluate the development and utilization of new energy. At present, China has not studied several types of new energy sources [6] . Only a few scholars have made a comprehensive evaluation of a new type of energy technology.

To sum up, this paper studies the patent technology of the three types of new energy, analyzes the trend of technology and import countries, regional technical domain analysis, trend analysis in technology field, especially the use of technology life cycle analysis. From the above analysis, this paper is to observe the development of China's three major renewable energy technologies after the renewable energy legislation in 2005. According to the development of the three major renewable energy technologies, this paper analyzes the shortcomings and enlightenment of the current Chinese renewable energy law in three major renewable energy technologies.

ANALYSIS AND RESEARCH

Overall Situation of Renewable Energy

As shown in Figure 1, The general trend of patent application for renewable energy is rise. There are 21 patents in 2008; There are 37 patents in 2009; There are 57 patents in 2010; There are 61 patents in 2011; There are 78 patents in 2012; There are 66 patents in 2013; There are 58 patents in 2014; There are 89 patents in 2015; There are 90 patents in 2016; There are 80 patents in 2017.

As shown in Figure 2, The field of patent application for renewable energy are eight fields. There are 292 patents in H; There are 238 patents in F; There are 138 patents in C; There are 83 patents in G; There are 80 patents in B; There are 30 patents in E; There are 17 patents in A; There are 2 patents in D. H is electricity; F are mechanical engineering, lighting, heating, weapons and blasting; C are chemistry and metallurgy; G is Physics, B are Work and transport; E is fixed building; A is Agriculture; D are textile and paper making.

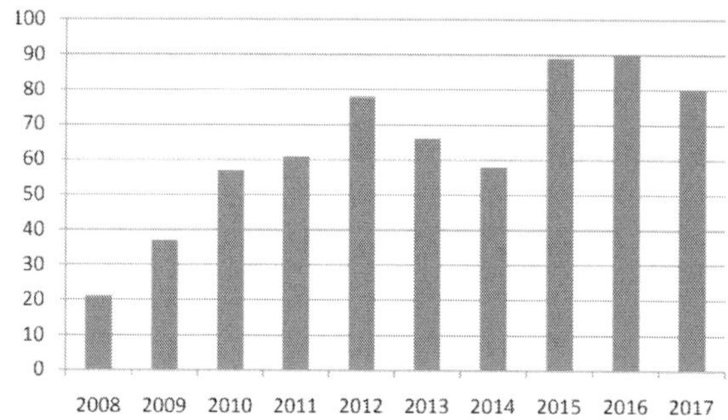

Figure 1. Trend of patent application for renewable energy.

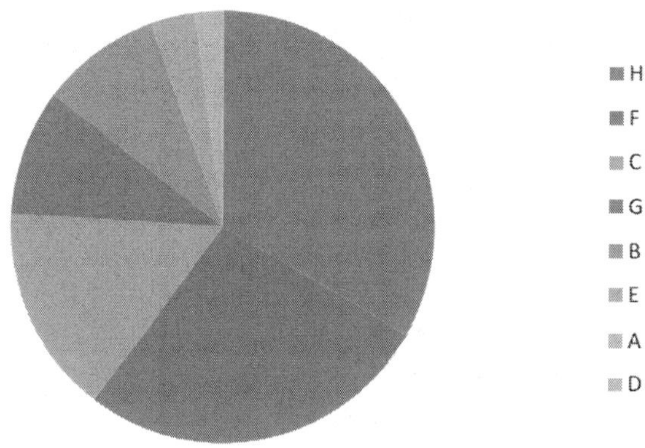

Figure 2. Application field of renewable energy patent.

Table 1. IPC ranking.

IPC	Number
H02	255
F03	100
F24	93
C10	57
G06	46
C12	43

As shown in Table 1, The first is H02 that are power generation, substation or distribution, there are 255 patents in H02; The second is F03 that are hydraulic, hydraulic engine, wind, elastic, gravity engine, other categories do not include mechanical or thrust engines, there are 100 patents in F03; The third is F24 that are heating, the stove or the ventilation, there are 93 patents in F24; The fourth is C10 that are oil, gas, coking industry, industrial gases containing carbon monoxide, fuels, lubricants or peat, there are 57 patents in C10; The fifth is G06 that are calculation, reckoning or counting, there are 46 patents in G06; The sixth is C12 that are biochemistry, beer, strong wine, juice wine, vinegar, microbiology, enzymology, mutation or genetic engineering, there are 43 patents in C12.

Nuclear Energy

As shown in Figure 3, most of China's nuclear energy technology is China's technology, and the main importing countries of nuclear technology are the United States, Japan, Germany, France and Russia. The import of nuclear technology in the United States imported technology accounted for 7.82%, Japan accounted for 6.41% of imported technology, Germany imported technology accounted for 4.46%, French imports of technology accounted for 1.67%, accounted for 0.55% of Russia's imports of technology.

As shown in Figure 4, G21 Technology: The main technology is China's autonomous technology, some imported from other countries, the number of importing countries is the largest of all statistical technology; F01 Technology: There is no import countries, all technology are China's autonomous technology; G06 Technology: In addition to China's independent technology, imports mainly from the United States and Taiwan; A62 technology is imported from Germany; E21 technology is imported from France.

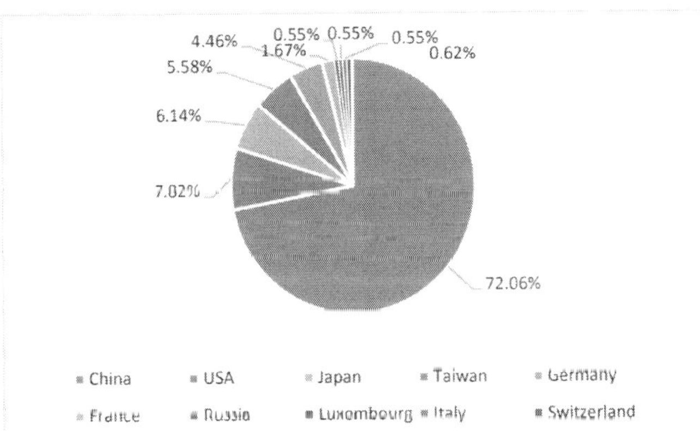

Figure 3. Regional map of nuclear energy technology.

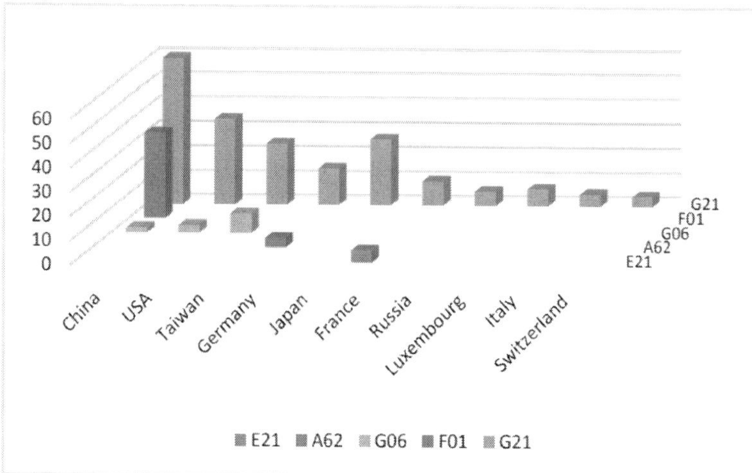

Figure 4. Domain regional analysis of nuclear energy technology.

As shown in Figure 5, The development trend of the 8 technologies in the period from 2008 to 2017 is in an unstable state. Generally speaking, G21 is in the rising stage.

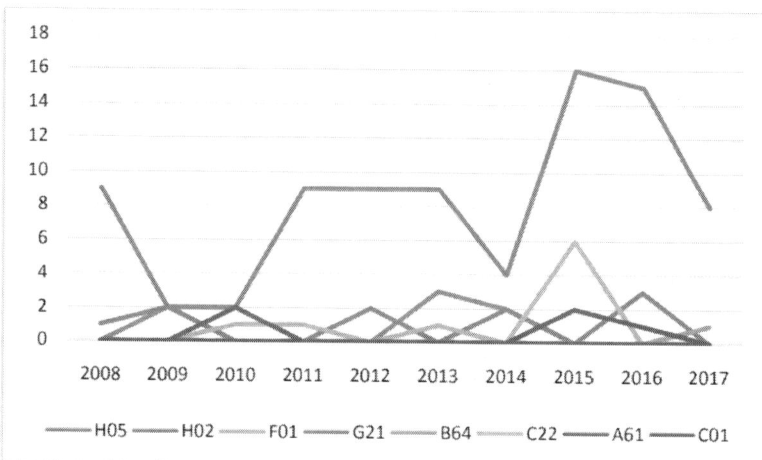

Figure 5. Trend analysis in nuclear energy technology field.

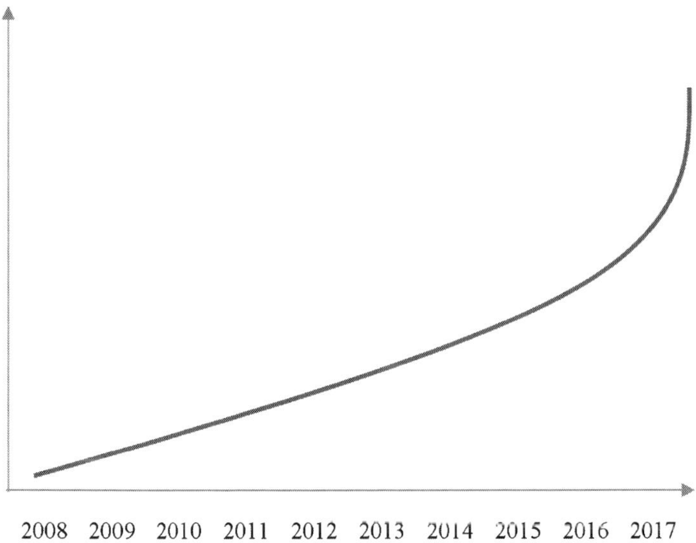

Figure 6. Life cycle of nuclear energy technology.

As shown in Figure 6, Almost all of the curves are in two stages: the number of patent applications and the number of applicants are relatively small, which is the initial stage of technology; and the number of patent applications and the number of applicants are increasing, which is the development stage of Technology. Only in these two stages, the technology start stage is more than the technology development stage.

Wind Energy

As shown in Figure 7, most of China's wind energy technology is China's technology, and the main importing countries of nuclear technology are Germany, USA, Canada, Denmark, Austria, Italy. The import of nuclear technology in the United States imported technology accounted for 1.07%, Germany accounted for 6.11% of imported technology, Canada imported technology accounted for 0.18%, Denmark imports of technology accounted for 0.16%, Austria imports of technology accounted for 0.16%, accounted for 0.16% of Italy's imports of technology.

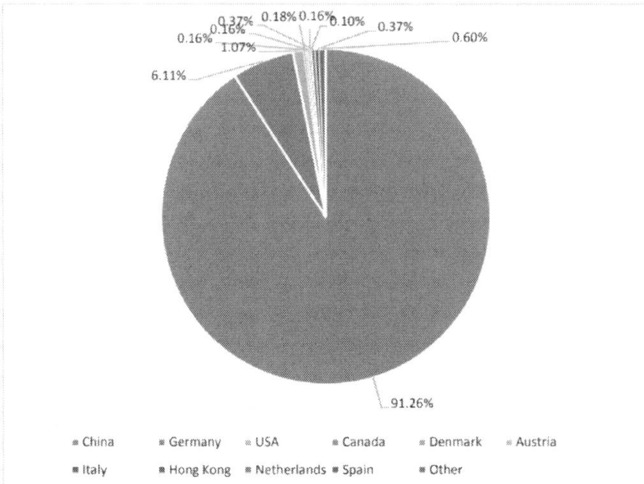

Figure 7. Regional map of wind energy technology.

As shown in Figure 8, F03 Technology: The main technology is China's autonomous technology, some imported from other countries, the number of importing countries is the largest of all statistical technology; H02 Technology: The main technology is China's autonomous technology, some imported from other countries, the number of importing countries is the second of all statistical technology; B60 Technology: In addition to China's independent technology, imports mainly from Germany, Canada; B63 technology: In addition to China's independent technology, imports mainly from Germany, Spain; F16 Technology: The main technology is China's autonomous technology, some imported from other countries, the number of importing countries is the third of all statistical technology; B09 technology: In addition to China's independent technology, imports mainly from Canada.

As shown in Figure 9, The development trend of the 2 technologies in the period from 2008 to 2017 are in the rising stage. But they are down in the last year.

As shown in Figure 10, Almost all of the curves are in one stages: the number of patent applications and the number of applicants are increasing, which is the development stage of Technology.

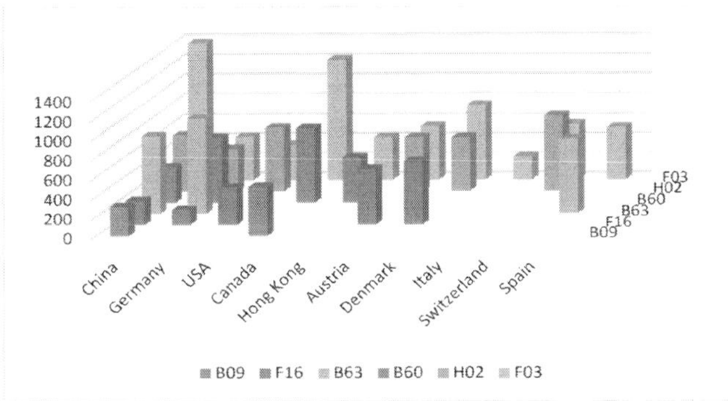

Figure 8. Domain regional analysis of wind energy technology.

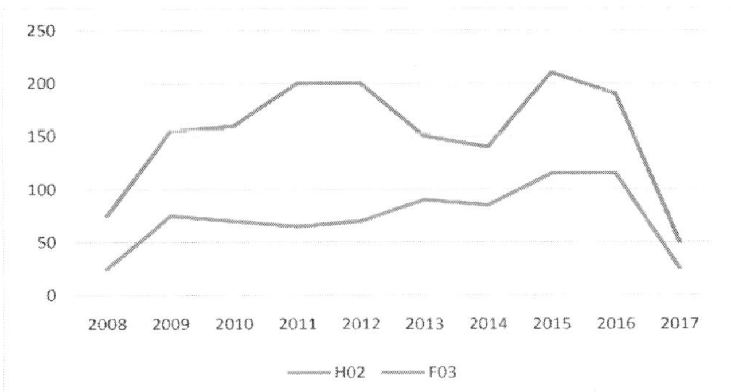

Figure 9. Trend analysis in wind energy technology field.

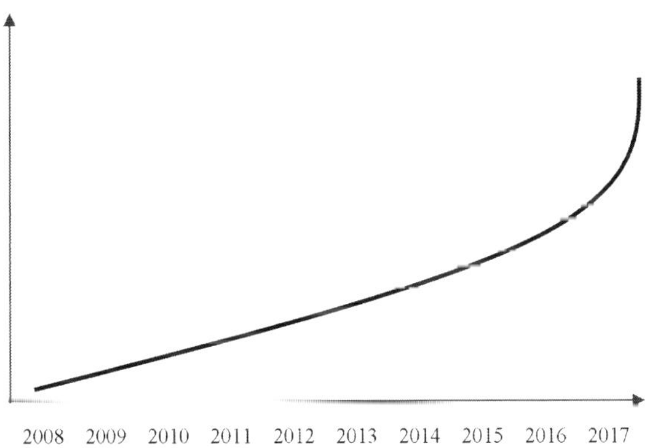

Figure 10. Life cycle of wind energy technology.

Biomass Energy

As shown in Figure 11, most of China's wind energy technology is China's technology, and the main importing countries of nuclear technology are Korea, Canada and Australia. The import of nuclear technology in Korea imported technology accounted for 0.74%, Canada accounted for 0.49% of imported technology, Australia imported technology accounted for 0.25%.

As shown in Figure 12, B09 technology: In addition to China's independent technology, imports mainly from Korea, Canada; F03 Technology: In addition to China's independent technology, imports mainly from Canada; F22 Technology: In addition to China's independent technology, imports mainly from Canada; C10 technology: In addition to China's independent technology, imports mainly from Korea, Australia; F24 technology: In addition to China's independent technology, imports mainly from Korea; C02, F22, C12, F23 Technology: There is no import countries, all technology are China's autonomous technology.

As shown in Figure 13, The development trend of the 4 technologies in the period from 2008 to 2017 is in an unstable state.

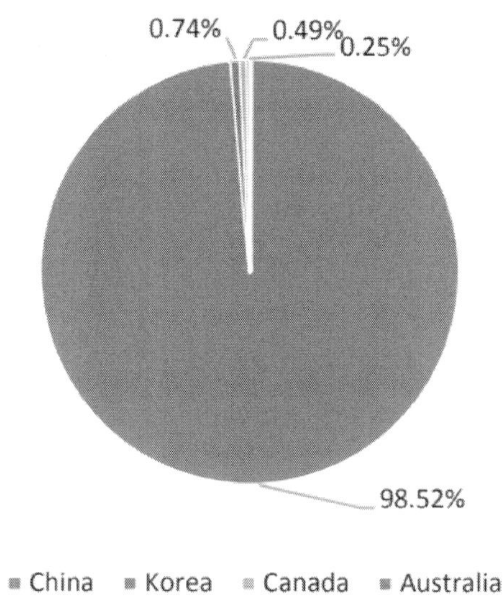

Figure 11. Regional map of biomass energy technology.

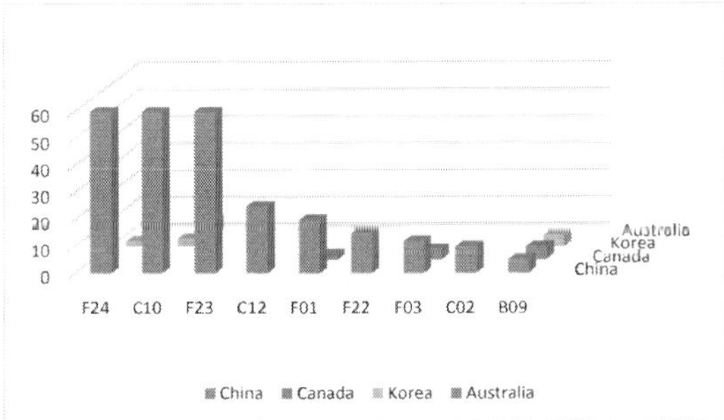

Figure 12. Domain regional analysis of biomass energy technology.

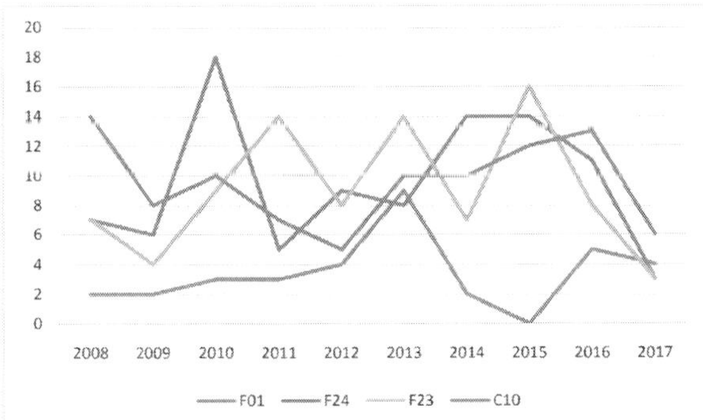

Figure 13. Trend analysis in biomass energy technology field.

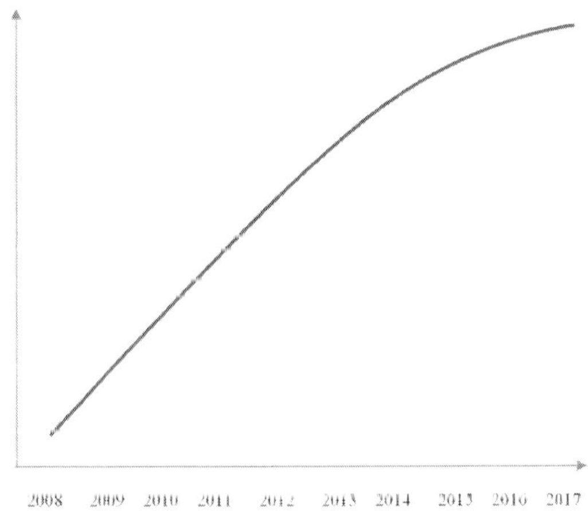

Figure 14. Life cycle of biomass energy technology.

As shown in Figure 14, Almost all of the curves are in one stages: the number of patent applications and the number of applicants are increasing, which is the development stage of Technology.

CONCLUSIONS

After the renewable energy law was approved in 2005, it has played a very important role in speeding up the development and utilization of renewable energy in China. It has not only become an important legal guarantee to promote the development of renewable energy in China, but also has a positive impact on the international community. From the above analysis, we can see that technology life cycle of China's three major renewable energy is in the growth stage. Thus, China's renewable energy law plays an obvious role [7] . The development of renewable energy is a long process in China, from the above analysis we can see three major shortcomings: The above the three renewable energy sources, some of which rely on imports in china; three major technical field development of renewable energy is uneven; The direction of technology development on three renewable energy is not obvious. The shortcomings of China's renewable energy development are summarized from the above analysis, and there are many inspirations for China's renewable energy policy: Firstly, China needs to support the import of renewable energy technology into autonomous technology, and gives more encouragement to the innovation of independent technology; Secondly, China needs to make clear the overall development direction of renewable energy technology, so that the development of technology and economy in renewable energy field can be controlled; Finally, China needs to balance the development of renewable energy technology in various fields according to the policy [8] . According to the above renewable energy policy, not only can accelerate the pace of China's renewable energy development, but also can make China's renewable energy technology development by macro-control and industry and field control.

REFERENCES

1. Chen, L.Q. (2010) Low Carbon Economy: A New Trend of Global Economic Development. Journal of Hunan City University, 31, 46-52.
2. Zhang, P. (2010) On the Response of Intellectual Property System to Low Carbon Technology Innovation. Science Technology and Law, 85, 29-32.
3. Huang, D. (2010) Low Carbon Technology Innovation and Policy Support. China Science and Technology Forum, 2, 37-40.
4. Zhao, Z. and Xiao, D.P. (2010) Bottlenecks and Countermeasures of Technological Innovation in Developing Low Carbon Economy. China Science and Technology Forum, 6, 41-46.
5. Chen, X.C., Hu, T. and Tang, Y.J. (2010) Discuss on Low-Carbon Consumption. Consumer Economics, 26, 83-85.
6. Wu, C.H. (2010) Roadmap of Technology Development for Low Carbon Innovation. Periodical of Chinese Academy of Sciences, 25, 138-145.
7. Langniss, O. and Wiser, R. (2003) The Renewables Portfolio Standard in Texas: An Early Assessment. Energy policy, 31, 527-35. https://doi.org/10.1016/S0301-4215(02)00095-2
8. Gireesh, S. and Joshua, K. (2011) Are Government Policies Effective in Promoting Deployment of Renewable Electricity Resource. Energy Policy, 39, 4726-4741. https://doi.org/10.1016/j.enpol.2011.06.055

SECTION 4:
CASE STUDIES FROM DIFFERENT COUNTRIES

14

A GIS Methodology for Planning Sustainable Renewable Energy Deployment in Portugal

Paula Costa, Teresa Simões, Ana Estanqueiro

Laboratório Nacional de Energia e Geologia—LNEG, Lisboa, Portugal

ABSTRACT

A Geographical Information System (GIS) methodology was developed to identify and characterize suitable areas for deploying renewable energy projects. The methodology enables to compute the sustainable renewable energy potential in an area under study and can be implemented for different spatial scales, ranging from local to national levels, while operating with different restriction layers. This GIS-based method has been successfully applied to wind energy deployment studies in Continental Portugal and other foreign countries (e.g. Venezuela, Mozambique among others). Results from several development plans using this methodology enable to conclude it is an adequate tool for planning sustainable renewable energy deployment both for onshore and offshore regions.

Citation: Costa, P. , Simões, T. and Estanqueiro, A. (2019), "A GIS Methodology for Planning Sustainable Renewable Energy Deployment in Portugal". Energy and Power Engineering, 11, 379-391. doi: 10.4236/epe.2019.1112025.

Keywords: Planning, Sustainable Potential, GIS, Renewable Energy, Wind Energy Deployment

INTRODUCTION

During the latest years, Portugal had a remarkable growth in the deployment of renewable energies (RE) due to the adoption of favorable legislation framework by the Portuguese government under the European Union energy policy. During 2013 the Portuguese government approved the new National Renewable Energy Action Plan [1] establishing a threshold of 5273 MW (by the year 2020) for the onshore installed wind power. As in other European Member states, Portugal is developing its National Plan for Energy and Climate (PNEC2030) [2], following the new European Commission Directive for Renewable Energy—Recast to 2030 (RED II) [3], and establishes very ambitious targets for renewable energy deployments (for wind energy, 8-9 GW, approximately, are foreseen until 2030). New and updated methodologies together with new regulations are though needed to comply with the established targets.

By the end of 2013, the total installed capacity in Portugal reached 4707 MW (23% of the total capacity installed in Portugal) [4] while in the end of 2017 the total installed capacity reaches 5313 MW [5] with a wind energy yearly penetration of 24% only surpassed by Denmark [5]. Although Portugal has acceptably good endogenous resources for exploiting REs the areas still available for new deployments are, becoming scarce therefore tools to identify available locations with good energy indicators are necessary.

In order to support policy makers and investors several GIS planning tools have been developed in recent years for assessing the sustainable potential of REs for several countries e.g. Spain [6] and US [7]. In addition, these planning methodologies were applied at regional scales [8] [9]. All these models take into account the various land uses, the wind resource—data from mesoscale modeling—and the environmental restrictions. In this line of work, LNEG's activity started in the early 90's with the development of the first wind energy database in Portugal, EOLOS [10] followed by EOLOS-2 [11]. These databases were developed in a Microsoft Access platform with SQL language with the ability to perform simple queries to the locally geo-referenced wind information. These databases do not provide or enabled spatial operations between data.

More recently, building on the historic knowledge of the REs resources over Portugal, the research goals were widened to include the spatial mapping of the onshore wind resource and the first Portuguese Wind Atlas [12] [13] [14] [15] and the Offshore Wind Atlas in 2006 [16]. Although the published studies were more focused on the development of wind resource atlases and therefore they do not include crucial information for planning purposes (e.g. physical and geographic constraints).

In this sense GIS studies were developed to obtain the sustainable availability of the renewable wind potential in Portuguese territory for the onshore case [17] and for the offshore case [18]. Section 2 of the paper provides a brief background of the REs methodology developed. Section 3 presents the results obtained with the methodology proposed for application to the wind energy Portuguese case study. Along the Section 4 a description of the GIS operation is provided and Section 5 presents the calculation of the sustainable REs potential for the wind energy case study. Finally, in Section 6 some conclusions are drawn.

METHODOLOGY

The proposed methodology requires the use of a GIS platform able to combine the use of different spatial operations taking into account the layers with constraints information and the renewable energy resource map. The methodology is divided in four sub-models applied sequentially. The first sub-model, inputs the constraint layers (environment, terrain slope, wind speed and wind energy resoure map) and compute the sustainable area which are the "free" areas adequate for wind park deployment. The second sub-model computes the soil occupation factors. The soil occupation factor corresponds to the normalized value of the number of inhabitants per kilometer squared per each administrative region. The third model computes for each area the total installed wind park capacity operating per each administrative region and the last model, computes the sustainable available potential per each administrative region. The output of this GIS methodology enables to identify the: a) best areas and the regions most adequate to energy deployments; and b) total sustainable renewable potential still available for the identified zones, according to the input premises classified onto to the restriction areas. Although the methodology can be applied to any spatially variable form of RE, this paper focuses on its development for the wind energy sector, using mainland

Renewable Energy Resource Definition

The base information for the application of the methodology is the development of (or assessment to) a geo-referenced resource map, usually defined as a resource atlas. This map constitutes the source information to which the exclusion criteria—associated with the n-dimensional layers of constraints—are applied thus assessing the final sustainable potential for each renewable form of energy. The resource map needs to reflect the economic sustainability of deploying a specific RE technology, thus an effective straightforward approach is to define a threshold condition that guarantees a minimum economic profitability of the technology under study, e.g. its annual equivalent number of hours at full capacity for wind turbines (NEPs). A database with the coordinates and nominal power of all wind power plants already operating (or under project) is also required and included in an additional GIS-layer. Table 1 shows the information required for input into the GIS methodology.

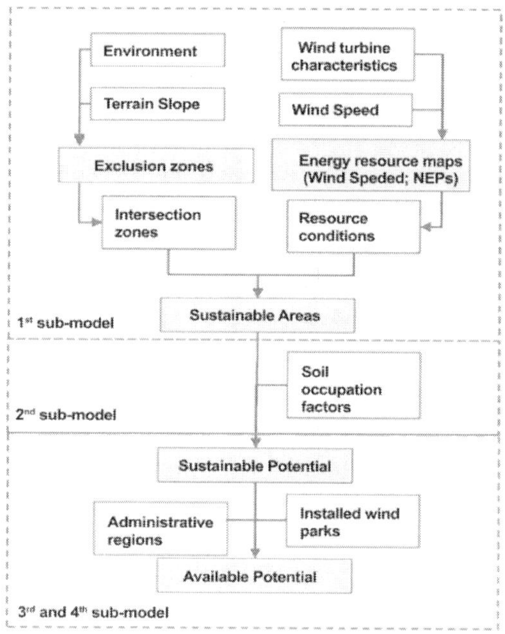

Figure 1. Methodology structure for the wind energy available potential assessment.

Table 1. Information required for GIS Methodology—renewable energy resource.

Information required for The GIS methodology:	• Energy resource map(s) from a renewable energy atlas • Minimum threshold condition value (e.g. Minimum mean wind speed (m/s) or NEPs (h/year)) • Reference wind turbine (WT) nominal power • Existing (under project) REs power plants • Equivalent renewable energy losses factor (MWh)

Criteria Classes and Spatial Definition Layers

To determine possible available areas, layers with human geography data such as administrative regions (districts, municipalities), roads and urban areas, among others, are ingested in-to the GIS. Physical geography information e.g. rivers, lagoons, terrain slope. Furthermore, the methodology incorporates information related to environment restrictions, such as, natural parks, protected heritage, and similar.

The regional social-economic occupation was taken into consideration. The occupation factors were defined according to the demographic classification of the area under study, e.g. number of inhabitants per km^2). The occupation factors are then classified from 0 to 1, where 0 means extremely dense occupation and 1 means extremely reduced populated areas. Table 2 shows the main spatial information and criteria classes for the GIS methodology application for the Portuguese onshore wind energy case study.

The methodology operates by integrating the information presented in Table 1 and Table 2 into a GIS platform. A set of models was programmed and organized in a designed toolbar that enables to identify available areas for the deployment of REs projects (wind energy, in the present case study) and, as a by-product to compute the sustainable REs capacity in each of those areas. In the current case study, the sustainable wind capacity can be mathematically expressed by the following expression:

$$P_i = \alpha\beta\gamma_i\eta_i\varepsilon_i - w_i \quad (1)$$

where, P_i represents the sustainable wind power (in MW) per identified area or polygon and The parameters involved in Equations (2) to (7) are expressed below:

$$\alpha = EL \quad (2)$$

$$\beta = \frac{Pot}{8760} \quad (3)$$

$$\gamma_i = SO_i \quad (4)$$

$$\eta_i = \frac{dxdy}{\delta x \delta y D^2} \quad (5)$$

$$\varepsilon_i = \sum_{j=1}^{N_i} NEPs_{ji} \quad (6)$$

$$w_i = \sum_{j=1}^{k} MW_{ji} \quad (7)$$

Table 2. Spatial information and criteria classes for GIS methodology.

Orography:	• Terrain slope
Environmental:	• Natural parks; Natura 2000 • Protected heritage
Soil occupation:	• Very low populated; low populated • Average populated; high populated • Very high populated

EL: coefficient for equivalent potential energy losses;

Pot: nominal power of a reference turbine (MW) used to compute the wind energy resource map in hours per year;

8760: the number of hours in a year;

SO_i: soil occupation factor for polygon i;

Dxdy: pixel area in the wind energy resource map (in meters);

$\delta x \delta y D^2$ land area required by each wind turbine, expressed as a multiple of rotor diameter D, where δx, δy represent the minimum distances for crosswind and along wind directions respectively (in meters);

$NEPs_{ji}$: NEPs map raster value after applying all GIS exclusions in grid point j inside polygon i;

N_i: number of grid points inside polygon i;

MW_{ji}: nominal power (MW) from each wind farm operating inside polygon i;

k: total number of wind farms inside polygon i.

THE PORTUGUESE WIND RESOURCE CASE STUDY

Resource Map: The Portuguese Wind Atlas

The resource map used as input for the Portuguese wind case study was obtained by numerical modeling of the atmospheric flow [12] [13] [14]. The resource map was computed at a spatial grid of 100 × 100 m having as reference a 2.0 MW nominal power wind turbine model. The wind energy resource map depicted in Figure 2 was processed for a standard height of 80 m above ground level and expressed in NEPs (h/year)

Terrain Slope

The terrain slope for the Portuguese territory was derived from a processed raster terrain database obtained by the Shuttle Radar Topography Mission SRTM30-Plus [19] [20]. For this case study, areas with slope values greater than 20% was exclude. Figure 3 shows the slope terrain in map in percent units used in the study.

Environment Restrictions

The Portuguese territory has specific legislation for areas inside the country that are protected and classified as "Natural Parks" and "Natura 2000" areas. Figure 4 depicts all these environmental areas.

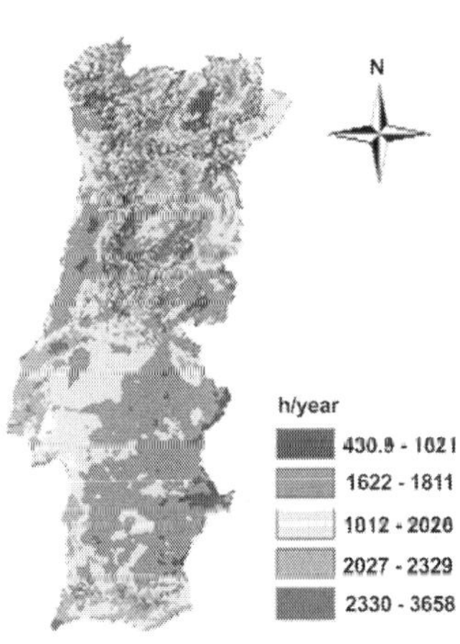

Figure 2. Spatial distribution of NEP's (h/year) over Portugal Mainland.

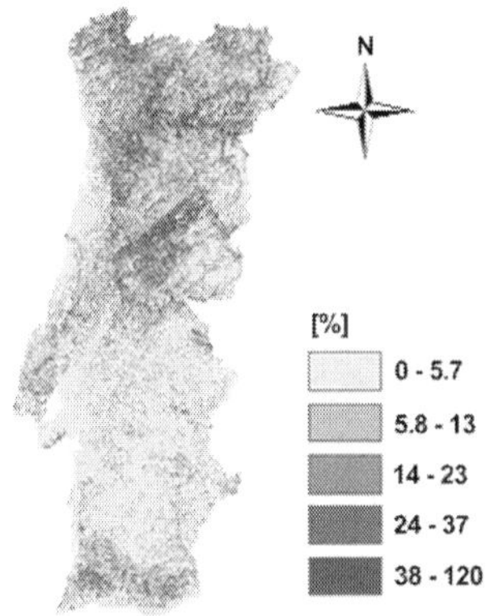

Figure 3. Raster map of the terrain slope (%).

Although these areas are classified as restricted there are some zones inside of them that may be used for wind energy exploitation with very low occupancy rates. For the case of "Natural Parks" an occupancy of 1% to 2% is assumed as acceptable. For the "Natura 2000" areas, and depending on the characteristics of each area, the occupation cannot exceed 25%.

Soil Occupation

The soil occupation should reflect the roughness of the terrain and the impact of the existing social economic activities. Roughness information can be provided by tabular values commonly used in wind engineering studies [21] [22] whose values were used in the previous wind energy resource assessment studies (§ 3.1 Wind Resource).

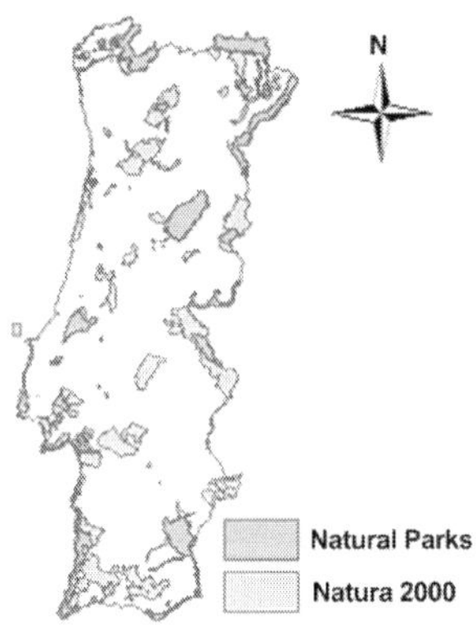

Figure 4. Environnemental areas (source http://www.icnf.pt).

For the Soil Occupation classification, a coefficient factor was defined for each administrative region depending on demography, namely the population occupation percentile, as in Table 3. The population occupation percentile was computed from the number of inhabitants per squared km per administrative area and the occupation factor corresponds to the normalized value of the number of habitants per squared km per each administrative region. In Figure 5 the soil occupation factors are depicted.

Active Installed Wind Parks

As part of the identification of the wind power capacity still available in each administrative region, it is mandatory to obtain information about the total wind capacity already installed. For the present case study, the most up to date information was gathered by municipality based in information provided both from DGEG institution (http://www.dgeg.pt) and by the annual publications of the International Energy Agency [3].

GIS APPLICATION

Model Development

The model development for the integration of the methodology into the GIS platform was performed with the help of Model Builder tool (ArcGIS® 10.0.4, ESRI software). This tool operates mainly over raster information (raster-based model) and enables the programming of spatial operations (union, intersection, clipping features or even mathematical expressions applied between layers), in an automatic form reducing the calculation time over the geo-referenced maps, which may be very long when large territories are under analysis. To perform the calculations a set of sub-models with the necessary operations and transformations was created.

Table 3. Spatial information and criteria classes for GIS methodology.

Soil occupation classification	Factor	Population percentile (%)
Very low populated	0.5	40
Low populated	0.4	60
Average populated	0.3	80
High populated	0.2	95
Very high populated	0.1	99

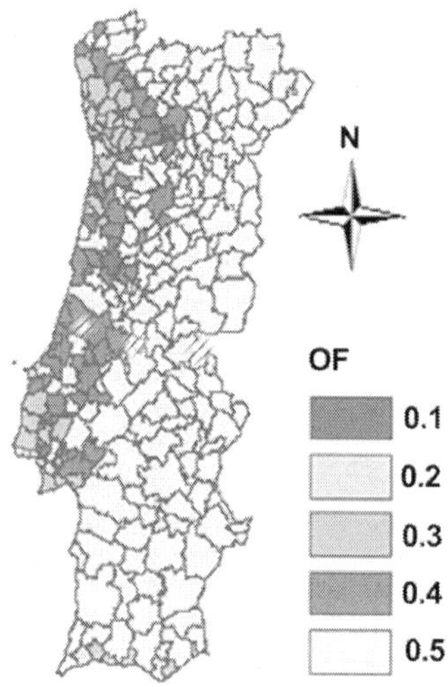

Figure 5. Spatial representation of the soil occupation classification (occupation factor).

The first sub-model treats for the operations concerned with restrictions between layers. This sub-model named "Define Restrictions" enables the definition of the restrictive layers and merges them into a raster dataset. This sub-model is also able to compute the terrain slope map and resample it according to the spatial resolutions of both wind energy and wind resource datasets, based on a digital elevation map. For the present case, the terrain slope, the Natural parks and "Natura 2000" areas were considered and conditions referred in 3.3 were applied. The output of this sub-model is used as an input information for the second sub-model. The second sub-model named "sustainable areas" refers to the application of the limits related to mean wind speed and wind energy maps according to the conditions described in 3.1. The third and fourth sub-models ("Sustainable Potential" and "Available Potential") are based on mathematical operations which compute the Equations (1) to (7). In particular the third sub-model, uses the soil occupation factors presented in 3.4 classified by municipality to obtain the sustainable wind potential and wind capacity still available for each municipality. Finally, the available wind potential is based on the sustainable wind potential obtained from the previous phase and on the installed wind capacity for each municipality.

Toolbar Development

In order to organize the calculations and make the procedures more efficient and less time consuming, a toolbar was developed containing all the sub-models. Therefore, the user is able to change the input data and criteria at any time and maintain, if necessary, the remaining calculations without changes, according to the objectives of the study. The procedures enable the identification of suitable areas for wind energy projects development, sustainable and available potential.

ASSESSMENT OF THE SUSTAINABLE WIND POTENTIAL

In this section the results provided by the toolbar "Wind Energy Planning" are presented for the Portuguese territory. Table 4 presents the values attributed to the input parameters for the current application, following Equations (1) to (5). Figure 6(a) depicts the results obtained with the first sub-model "Define Restrictions" where the terrain's slope and the environmental restriction layers (Natural Parks and Natura 2000 areas) were used according to the assumed conditional information presented in Table 4 and Figure 6(b) shows the results from the second sub-model, "Sustainable Areas", where the outputs from the first sub-model were ingested and the conditional information from the mean wind speed, NEPs, and the environmental restrictions were applied.

The next step comes from the sub-model "Sustainable Potential". In this case the sub-model operates over the map presented in Figure 6(b) and applies to the restrictions expressed by the mathematical formulation according to Equations (1) to (7). For regional planning purposes, the method includes the municipalities' layer, to assess adequate areas for RE projects deployment inside each administrative region. It should be noted that soil occupation factors according to Figure 5 are applied in this step. Finally, the available potential per municipality can be estimated using the last sub-model "Available Potential". Figure 7 presents the available potential for wind park deployment for each municipally. In this last sub-model, information about the already installed wind capacity in Portugal was used and Table 5 represents the results referred to the mainland territory considering the available information about the Portuguese onshore total operating wind parks at the end of 2017.

Table 4. Parameter values used in GIS Methodology according to Equations (1) to (5).

Parameter	Value
EL (energy losses)	0.95
Pot	2.00
D	80.00
dx, dy	100.00
δx, δy	5.00
NEPs (h/year)>	2100.00
Mean wind speed (m/s)>	6.00
Terrain slope (%)<	20.00
Release natural park area (%)	2.00
Release natural 2000 area (%)	25.00

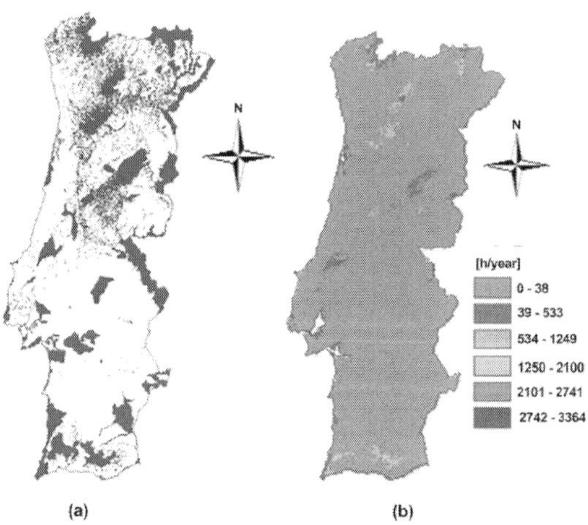

Figure 6. (a) Map with the environment and slope restrictions. (b) Suitable areas for wind park deployment over layed with NEPs values (h/year).

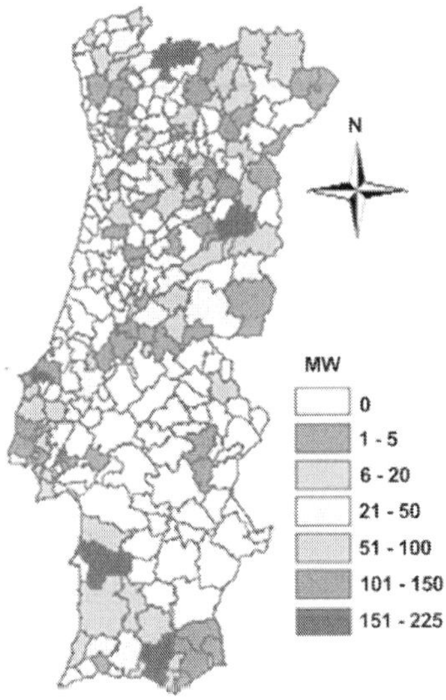

Figure 7. Available potential per municipality (MW).

Table 5. Sustainable available wind energy potential in continental Portugal (in MW).

Sustainable Wind Potential (MW)	NREAP Targets (2020) (MW)	Total Install Wind Cap. 2017 (MW)	Available Wind Pot. (MW)
6428	5273	5313	1115

The results from Table 5 enable to conclude that the sustainable wind energy in Portugal is nearly 22% upper than the fixed target in NREAP 2013 for 2020. Actually, the total installed capacity in the Country has overpassed the NREAP target and according the obtained results the available wind potential in the Mainland is circa 1115 MW still available for newer wind power projects.

CONCLUSION

In this paper a methodology for the identification and quantification of the sustainable renewable potential using geographical information systems (GIS) was presented. The methodology was developed as a flexible tool to allow energy planners to test and alter RES conditions and restrictions in a simple and straightforward manner. The methodology was programmed and organized in a "toolbar" which enables the user to execute the sub-models developed for each phase of the planning process. The methodology is applied to the assessment of the available wind potential assessment in continental Portugal. This case study allows to illustrate the type of results that can be obtained and the added value of such a tool. Within a framework of sustainable development of the renewable energy sector, the concept and methodology presented can be applied to any geographic region where reliable information exists such as the wind resource map, geographic, environment and societal restrictions as minimum requirements to this methodology to evaluate successfully the sustainable wind potential for any region. Facing the new challenges imposed by the targets established in PNEC2030, new simulation with higher resolution and considering the repowering of old wind turbines, among other considerations, is undergoing.

ACKNOWLEDGEMENTS

The authors want to acknowledge to the Direção Geral de Energia e Geologia (DGEG) to provide the total wind capacity installed per administrative region for the year of 2017 and to the Instituto Nacional de Estatística (INE) and PORDATA website for providing relevant statistical data about the number of inhabitants per km squared for the 2017 year.

REFERENCES

1. (2013) National Renewable Energy Action Plan-NREAP. http://ec.europa.eu/energy/efficiency/eed/doc/reporting/2013/pt_2013report_en.pdf
2. (2018) Plano Nacional Integrado Energia E Clima 2021-2030. https://ec.europa.eu/energy/sites/ener/files/documents/portugal_draftnecp.pdf
3. Directive (EU) 2018/2001 of the European Parliament and of the Council of 11 December 2018 on the Promotion of the Use of Energy from Renewable Sources (Recast). https://eur-lex.europa.eu/legal-content/EN/TXT/PDF/?uri=CELEX:32018L2001&fr om=EN
4. International Energy Agency (2013) IEA Wind 2013 Annual Report.
5. IEA Wind TCP Annual Report 2017 for Portugal. https://community.ieawind.org/publications/ar
6. Bravo, J., Casals, X. and Pascua, I. (2007) GIS Approach to the Definition of Capacity and Generation Ceilings of Renewable Energy Technologies. Energy Policy, 35, 4879-4892. https://doi.org/10.1016/j.enpol.2007.04.025
7. Lopez, A., Roberts, B., Heimiller, D., Blair, N. and Porro, G. (2012) U.S. Renewable Energy Technical Potentials: A GIS-Based Analysis. Technical Report. https://doi.org/10.2172/1219777
8. Simões, T., Costa, P. and Estanqueiro, A. (2002) Desenvolvimento de Mapas de Vento em Portugal-Aplicação da Metodologia à Região Algarvia. Portuguese at XI Congresso Ibérico e VI Congresso Ibero-Americano de Energia Solar, Vilamoura, Portugal.
9. Grassi, S., Chokani, N. and Abhari, R. (2012) Large Scale Technical and Economical Assessment of Wind Energy Potential with a GIS Tool: Case Study Iowa. Energy Policy, 45, 73-85. https://doi.org/10.1016/j.enpol.2012.01.061
10. Simões, T. and Estanqueiro, A. (2000) Base de Dados do Potencial Energético do Vento em Portugal-EOLOS, CD-ROM Publication, INETI/DER.
11. Brandão, R, Rio, J., Costa, P., Teixeira, J., Simões, T and Estanqueiro, A. (2004) Base de Dados do Potencial Energético do Vento em Portugal-EOLOS 2.0. CD-ROM Publication, INETI/DER.
12. Costa, P. (2004) Atlas do Potencial Eólico para Portugal Continental. Master Degree, Faculty of Sciences, University of Lisbon, Portugal.

13. Costa, P. and Estanqueiro, A. (2004) Atlas do Potencial Eólico para Portugal Continental. CD-ROM Publication, INETI/DER.
14. Costa, P., Miranda, P. and Estanqueiro, A. (2006) Development and Validation of the Portuguese Wind Atlas. Publication in the Proceedings of the European Wind Energy Conference (EWEC), Athens.
15. Costa, P., Miranda, P. and Estanqueiro, A. (2006) Validation of the Portuguese Wind Atlas. Conference in Métodos Numéricos em Mecânica dos Fluidos e Termodinâmica, Faculty of Sciences and Technology, Monte da Caparica, Portugal.
16. Costa, P. and Estanqueiro, A. (2006) Assessment of the Sustainable Offshore Wind Potential in Portugal. Publication in the Proceedings of the European Wind Energy Conference (EWEC), Athens.
17. Simões, T., Costa, P. and Estanqueiro, A. (2008) A Methodology for the Identification of the Sustainable Wind Potential. The Portuguese Case Study. 2009 IEEE/PES Power Systems Conference and Exposition, Seattle, WA, 15-18 March 2009, 1-7. https://doi.org/10.1109/PSCE.2009.4839951
18. Costa, P., Simões, T. and Estanqueiro, A. (2010) Sustainable Offshore Wind Potential in Continental Portugal. Proceedings of the Workshop "Oceans as a Source of energy", Academia de Engenharia, Berlin-Brandenburgische der Wissenschaften, Lisbon, Portugal, 40-43
19. Farr, T.G., et al. (2007) The Shuttle Radar Topography Mission. Reviews of Geophysics, 45, RG2004. https://doi.org/10.1029/2005RG000183
20. Rodriguez, E., Morris, C.S., Belz, J.E., Chapin, E.C., Martin, J.M., Daffer, W. and Hensley, S. (2005) An Assessment of the SRTM Topographic Products. Technical Report JPL D-31639, Jet Propulsion Laboratory, Pasadena, California, 143 p.
21. Simiu, E. and Petersen, R.H. (1986) Wind Effects on Structures. Wiley Interscience, New York.
22. Troen, I. and Petersen, E.L. (1989) European Wind Atlas. Riso National Laboratory.

15

The Necessity of the Development of Standards for Renewable Energy Technologies in Nigeria

Vincent Nnaemeka Emodi, Samson D. Yusuf, Kyun-Jin Boo

Department of Technology Management, Economics and Policy Program, College of Engineering, Seoul National University, Seoul, South Korea

ABSTRACT

Clean energy is vital for the sustainability of any economy in the world. Many industrialized nations have increased their production capacity of renewable energy while other countries lacking the technical expertise and resource have resorted to import these technologies. The imported technologies mostly have standards that are followed by the manufactures while others are manufactured cheaply and exported to developing countries that do not have adequate standards and certification bodies. Nigeria which is a fast growing country has no existing standards to check the influx of renewable energy technologies into the country. This study stresses the need for the development of standards for renewable energy technologies in order

Citation: Emodi, V, Yusuf, S. and Boo, K. (2014), "The Necessity of the Development of Standards for Renewable Energy Technologies in Nigeria". Smart Grid and Renewable Energy, 5, 259-274. doi: 10.4236/sgre.2014.511024.

to prevent the importation of substandard renewable energy technologies in Nigeria. The study reviews the renewable energy potentials in Nigeria, introduces the concept of standardization and discusses the development of standards for renewable energy technologies.

Keywords: Standards, Standardization, Renewable Energy Technologies, Substandard

INTRODUCTION

Energy is an important factor for socioeconomic development and economic growth of any nation. Energy is usually stored in energy system which provides energy services. Energy services are desired and useful products, processes or services that result from the utilization of energy like lighting, powering of home-based appliances such as air-conditioner, refrigerators, and cookers for cooking. The energy chain to deliver these cited services begins with the collection or extraction of primary energy, which is converted to energy carriers suitable for the final consumers. These energy carries are utilized in energy end-user technologies to provide the desired energy services [1] .

Energy sources are divided into conventional energy which is crude oil, natural gas and coal while non-con- ventional energy or renewable energy is solar, wind, hydro and biomass energy [2] . About 90% of the total energy supply in the world comes from conventional energy. Conventional energy contains mostly carbon which is the leading cause of global warming when released from crude oil products, gas and coal to the atmosphere [3] . The increase in global warming has indulged both the developed and developing countries to make a shift to renewable energy resources.

Most energy projection shows that the current and future world energy demand patterns are definitely not sustainable. However, the demand for energy will increase dramatically with developing countries consuming more and more energy. The only solution is to decouple the economic activities associated with conventional energy consumption [4] . This can be achieved by increasing the share of renewable energy in the energy consumption balance to enhance sustainability and improve the security of energy supply [5] .

Renewable energy development in developing country aims at providing sustainable energy which will ensure the proper mitigation of energy scarcity and global warming. Renewable energy development has led to the increase in the manufacturing of renewable energy technologies such as solar panels, wind turbines and small hydropower (SHP). These technologies transform the energy stored in the sun, wind and hydro to energy which can be harnessed for human consumption.

Renewable energy that is being manufactured by different industries in various countries of the world has their specification and standards. When these technologies are imported into another country, standards are created by standards bodies in the importing country to ensure that the renewable energy technologies are reliable, compactable and safe for its citizens. Countries that don't have the technical expertise to manufacture renewable energy technologies have to import from other countries like China, South Korea, Japan, and the US etc. The countries that lack technical expertise are mostly in the African continent and they rely on imported technologies. These technologies are sometimes substandard and unreliable while others are even dangerous to the consumers and can cause fire outbreak which can destroy lives and property.

Nigeria, which is located in the West African region has been importing renewable energy technologies but has failed to ensure the establishment of standards for renewable energy by the Standards Organization of Nigeria (SON). This has led to the import of substandard renewable energy technologies into the country and has posed some dangers to her citizens who pay with their money hard earning but get low value for their money [6] .

The question that arises now is: what should be done to ensure that this menace of the influx of substandard renewable energy technologies into the country is checked and how can a renewable energy standard be developed in Nigeria. This study seeks to answer the research questions by first reviewing the renewable energy resources in Nigeria, identifying the barriers to its development, provides strategies to enhance its development and then discusses on the development of renewable energy standards. The remainder of this paper is organized as follows. Section 2 reviews the renewable energy potentials in Nigeria including the policies, barriers and strategies to enhance its development. The concept of standardization is given in Section 3 which also includes the roles of standards and the standards regulatory agency in Nigeria. Section 4 discusses the development of renewable energy standards in Nigeria while conclusions are provided in Section 5.

RENEWABLE ENERGY RESOURCES IN NIGERIA

The global mindfulness of the phenomenon of climate change along with the anticipation of conventional energy scarcity have prompted many countries to develop a more sustainable energy system to cater for economic development and growth. Clean and environmental friendly energy can only be achieved through the proper utilization of renewable energy technology [7] . The renewable energy resources in Nigeria are as enormous as they are diverse [8] . However, the problem lies with the level of utilization which is very low. A long term commitment from the Government of Nigeria (GON) is crucial in implementing any kind of policies which will lead to renewable energy development, this can be seen in countries like Germany, Denmark, Japan and more recently, South Korea [9] .

Nigeria's opportunity to improve the standard of living for its citizens, ensuring socioeconomic and political growth depends on the nation's ability to increase energy supply and proper utilization of its energy resources starting from the grassroots level [10] . Renewable energy technology has great potentials in alleviating the staggering energy situation currently being experienced in Nigeria. The potentials of renewable energy resources in Nigeria is about 1.5 times that of conventional energy resources in energy terms [11] . Table 1 [12] shows the renewable energy resources in Nigeria and they are discussed in the following subsections. It should be noted that biomass energy was not discussed in this study because biomass energy technology includes biomass plants and therefore this study only focused on solar, wind and hydropower.

Solar Energy

Among all the renewable energy resources available, solar is the most promising of them all due to its apparent limitless potentials. The energy that is radiated from the sun is about 3.8 × 1023 KW per second and most of it is transmitted radially as electromagnetic radiation which gives about 1.5 KW/m^2 at the boundary of the atmosphere, a square meter of the earth's surface can receive solar power of about 1.5 KW which averages about 0.5 daily [13] .

Nigeria is located within a high sunshine belt and solar radiation is well distributed within the country. The intensity of solar radiation exhibits remarkable variation from the northern region to the southern region but is higher in the northern region as shown in Figure 1 [14] . With an average domestic load demand of 2324 Wh/m^2 per day [15] , Nigeria has a solar radiation potential of about 6500 Wh/m^2 in the far north and 4000 Wh/m^2 in the southern part of the country. Given an average solar radiation level of about 5500 Wh/m^2, and prevailing efficiencies of commercial solar electric generators, then if solar collectors were employed to cover about 1% of Nigeria's land area of 923,773 KM, about 1850 × 103 GWh of solar electricity could be generated per year [16] [17] .

The Energy Commission of Nigeria (ECN) has made some effort to harness the solar energy within Nigeria through the direct coordination of research and development activities undertaken by the Sokoto Energy Research Center (SERC) and the National Center for Energy Research and Development (NCERD) [18] . However, this effort has seen little or no significant impact in improving the energy situation in Nigeria. There is need for a strong industrial infrastructure which will be effective in the complete utilization of solar energy in Nigeria.

Wind Energy

Wind is a natural phenomenon related to the movement of air masses caused primarily by the movement of air masses caused primarily by the different solar heating of the earth's surface. Seasoned variations in the energy received from the sun affects the strength and direction of the wind. The ease with which aero turbines transform energy in the moving air to rotary mechanical energy suggest that the use of electrical devices to convert wind energy to electricity. For decades, wind has been used for water pumping and milling of grains [1] .

Table 1. Renewable energy resource in Nigeria (source: [12]).

Resource type	Reserves		Production	Domestic utilization (natural units)
	Natural units	Energy units (Btoe)		
Small hydropower	3500 MW	0.34 (over 40 years)	30 MW	30 MW
Large hydropower	11,250 MW	0.8 (over 40 years)	1938 MW	1938 MW
Wind	2 - 4 m/s at 10 m height (mainland)	0.0003 (4 m/s@12% probability, 70 m height, 20 m rotor, 0.1% land area,40 years)	-	-

Solar radiation	3.5 - 7.0 kWh/m²/day (4.2 million MW h/day using 0.1% land area)	5.2 (40 years and 0.1% land area)	6 MW h/day	6 MW h/day
Biomass fuel wood	11 million hectares of forest and wood land excess of 1.2 m ton/day	-	0.120 million ton/day	0.120 million t/day
Animal waste	211 million assorted animals waste	-	0.781 million ton of waste/day	None
Energy crops and agricultural residue	28.2 million hectares of arable land (=30% of total land)	-	0.256 million ton of assorted crops/day	None

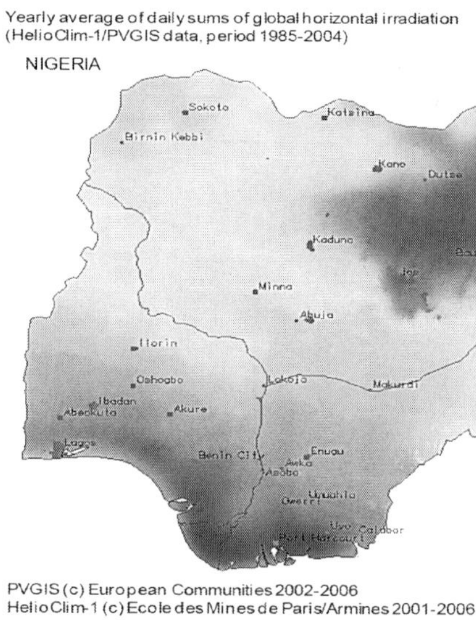

Figure 1. Yearly average of daily sun in Nigeria (source: [14]).

Wind speed in Nigeria ranges from 4.0 to 6.0 m/s in the northern part of the country and 2.5 to 4.0 m/s in the southern part of the country as shown in Figure 2 [19] . With this amount of wind energy potential, small scale wind turbine could be installed to boost electricity supply and also be integrated into the national grid [8] .

Despite Nigeria's exploitable wind energy resources, the present share of wind energy in the national energy consumption has been low with no available commercial wind farms but only standalone wind power plants for pumping water which were installed in 5 northern states during the 1960s. Also, a 10 MW wind turbine is currently installed in Katsina state to tap into the wind energy potentials in northern Nigeria. More effort should be put in to adequately harness the wind energy potential in Nigcria.

Hydropower

Hydropower generation is an important option to meet the growing demand for energy worldwide. In 2004, the world hydropower capacity was 2810 TWh and is projected to increase to 4903 TWh by the year 2030, with a growth rate of 1.8% per year, but its share still remains at 2% of the world energy supply [19] . Hydropower resources in Nigeria was first harnessed in 1962 by the Niger Dams Authority. Hydropower generation in Nigeria has substantial potential like the small, mini and micro water capacity for electricity generation but its total power contribution has declined in recent years due to some technical reasons [20] . Moreover, the use of oil and gas for power generation has had a negative impact on hydropower development in Nigeria [21] .

Notwithstanding, small hydropower (SHP) is fast gaining rapid consideration due to its inherent advantages like in-excessive topography problems, low environmental impact, minimal civil works and high possibility for power generation along with irrigation and flood prevention. Research has been conducted by [22] and shows that about 738 MW of SHP could be harnessed from 278 sites based on the 1980 survey. However, the SHP potential in Nigeria is estimated at 3500 MW which represent about 23% of the total hydropower potential in the county.

Hydro capacity depends mainly on annual rainfall levels, with its distribution as well as the river systems which is subject to seasonal drought. In the northern part of Nigeria, the total rainfall is about 500 mm depth with a total precipitation which last for 3 months in a year while the southern part of the country has 3400 mm with a precipitation which may be less than 8 months a year [23] . Figure 3 shows the various water ways in Nigeria [24] . The government should do more to develop SHP to its full extent in order to improve power generation and reduce fiscal loads.

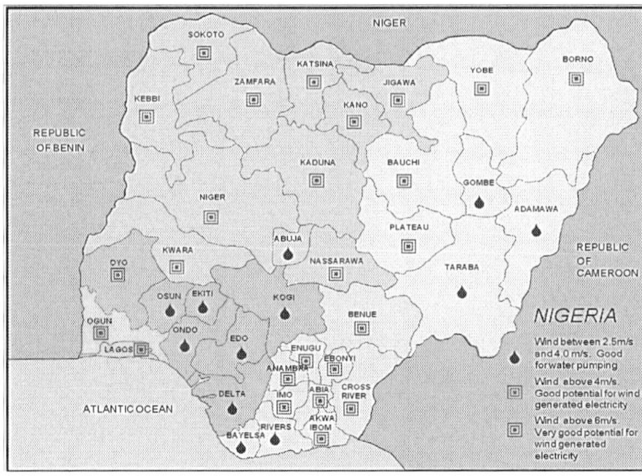

Figure 2. Wind energy locations in Nigeria (source: [19])

Figure 3. Nigerian water ways (source: [24]).

Renewable Energy Policy

The Energy Commission of Nigeria (ECN) has a mandate as an agency to develop and promote renewable energy technologies in Nigeria with strategic energy planning, policy coordination and performance monitoring for the entire energy sector. The ECN also has the responsibility of providing guidelines for the utilization of the various energy types for the different purpose and develop recommendations on new exploitable sources of energy. This includes renewable energy [25] .

The Federal Government of Nigeria (FGN) in 2003 approved the National Energy Policy (NEP) which was facilitated by the ECN to articulate the sustainable exploitation and utilization of all energy sources. The key elements of the NEP on the development and utilization of renewable energy and its technologies are as follows:

- To develop, promote and harness the renewable energy resources of Nigeria and incorporate all viable ones in the national energy mix;
- To promote decentralized energy supply, especially in rural areas, based on renewable energy resources;
- To deemphasize and discourage the use of wood as a fuel;
- To promote efficient methods in the use of biomass energy resources;
- To keep abreast of the international development in the renewable energy technologies and applications [26] .

In 2006, the ECN also formulated the Renewable Energy Master Plan (REMP) as part of the strategy to reduce Greenhouse Gas (GHG) emission in Africa and address the challenges hindering the development of clean, reliable, secured and competitive energy supply [27] [28] . The REMP main objectives are given below:

- To develop and implement strategies that will achieve a clean reliable energy supply mechanism to develop the sector based on international best practices, to show case viability for private sector participation;
- To provide a comprehensive framework for developing renewable energy that will ensure:

Ø Expanding access to energy services to Nigerians;

Ø National agenda on emission reduction;

Ø Raising the standard of living, especially in the rural areas;

Ø Stimulating economic growth, employment and empowerment;

Ø Increasing the scope and quality of rural services, including schools, information, health services, entertainment, and water supply and reducing rural urban migration [29] .

The REMP aims at a 10% renewable energy contribution to the national energy mix by 2020 through the adoption of a renewable portfolio standard (RPS). A RPS is a requirement for electric utilities to supply a specific amount of electricity to customers. This can be achieved through the purchase of renewable energy certificates from suppliers with a larger share of renewables in their energy mix. Other measures considered are the creation of innovative fiscal and market incentives to promote renewable energy industries, as well as preferential customs duty exemptions for imported renewable energy technology components. However, the lack of implementation of the master plan has meant that the 10% target of renewable energy mix in the energy supply cannot be achieved. The REMP is presently being subjected to a review, likely resulting in the setting of new targets. It will be essential that any future targets set for the attainment of a RE energy mix should be backed by legislation to ensure compliance, which is presently lacking.

Similarly, the National Policy and Guidelines on Renewable Electricity (NPGRE) was produced in 2006 with the main aim to expand the market for renewable electricity by 5% of the total electricity generation by 2016. The strategy for achieving this target included: encouraging local manufacture and assembly of renewable energy components, provision of subsidies, and establishment of technical standards for RE components and introduction of feed-in-tariffs. The strategy is yet to be fully adopted as the reforms in the energy sector are still ongoing and decisions on tariffs and subsidies for renewable energy and other incentives have not yet been taken [8] .

Barriers to Renewable Energy Development

Renewable energy in Nigeria has been inconsistent with minimal or no significant impact in the lives of Nigerians. Many barriers hinder the development and diffusion of renewable energy in Nigeria, including the following [30] :

- The lack of proper institutional framework which is due to the weak coordination between government ministries and agencies. These institutions are responsible for the promotion and development of renewable energy technologies in both the rural and urban areas;
- Low human capacity in the field of renewable energy development. Capacity building in the areas of training to install, operate, maintain and manufacturing of renewable energy is lacking in Nigeria;
- Non-existing Power Purchasing Agreements (PPA) plan for renewable energy generation to the utility grid. The PPA set the

terms in which electricity is marketed and determines the delivery location, power characteristics, pricing, quality, schedule and terms of agreements including the punishment for breach of contract;

- Affordability of renewable energy technologies involves high cost of capital as compared to conventional energy. This has posed a lot of problems to investors who face the problem of high transaction costs and restricted to funds or capital. Consumers on the other side also face difficulty in purchasing these technologies because of its expensive cost of purchase;
- The lack of public awareness has negatively affected the development of renewable energy in Nigeria. Most Nigerians view renewable energy technologies as new technologies that are only for the wealthy in the society. This has made the most of the population who are low income earners to depend on conventional energy which is cheaper;
- Uncertified/poor quality renewable energy technologies now flood the markets in Nigeria. This has led to the withdrawal of most potential consumers because most renewable technologies easily damaged due to their low quality and the replacement parts or services usually cost a lot to obtain. Nigeria presently lack the proper standards in renewable energy technologies.

Strategies to Enhance Renewable Energy Development

As Section 2.5 highlighted the barriers to renewable energy development, this section will offer some strategies which will promote renewable energy development in Nigeria:

- Placing subsides on renewable energy technologies and also improving access to micro finance loans by potential consumers;
- The government should do more in the area of public awareness campaign to promote renewable energy technologies;
- Research and development in the area of renewable energy should be encouraged. This will enable the growth of human capacity and ensure that the technologies which are mostly imported can be manufactured in the country;
- A legally binding long term Power Purchase Agreement plan should be established in order to encourage developers of

renewable energy. Also, other renewable energy development mechanisms like Feed-in-Tariffs (FiTs), Renewable Obligation (RO), Renewable Energy (green) Certificate (REC) and Renewable Portfolio Standard (RPS) should be initiated to facilitate the adoption of renewable energy technologies in Nigeria;

- Development of Renewable Energy Standards by the Standards Organization of Nigeria to check the influx of substandard renewable energy technologies into the country.

STANDARDIZATION

The process of establishing and applying standards, standardization is defined by the International Organization for Standardization as "the process of formulating and applying rules for an orderly approach to a specific activity for the benefit and with the cooperation of all concerned, and in particular for the promotion of optimum overall economy, taking due account of functional conditions and safety requirements". Standardization can be applied to specific products, as well as to, for example, norms, requirements, methods, terms, and designations commonly used in international trade and in science, engineering, industry, agriculture, construction, transportation, culture, public health, and other spheres of the national economy [31] . Standardization has a significant influence on the rate of development and level of production. Standardization provides the following advantages:

- Better product quality, reliable and durable life service;
- Components or parts mass production at a low cost;
- Availability of parts for replacement and maintenance;
- Less time and effort (productivity is high) for manufacture;
- Sizes and grades variations correction.

The concept of standardization uses numbers to limit unnecessary variations in sizes and grades of products, the general requirements indicates that such requirements are satisfied when it follows a geometric series. In international business, standards are applied to penetrate the markets in such a way that the products are made to fit into the requirements of any country. To do so, the need to follow the internationally recognized guidelines arises. Examples of products standard includes fuel economy and airbag specifications.

Definition of Standards

A "standard" can be defined as a specification (or a set of specifications) that relates to a product's attributes [32] . The International Organization for Standardization (ISO) defines standards as "a document, established by consensus and approved by a recognized body that provides, for common and repeated use, rules, guidelines or characteristics for activities or their results, aimed at the achievement of the optimum degree of order in a given context" [33] . In the WTO, definition for standards differentiates voluntary standards and technical regulation, where standards are voluntary and technical regulations mandatory with administrative provisions [34] .

The IRENA defines standards as "a repeatable, harmonized, agreed and documented way of doing something. Standards contain technical specifications or other precise criteria designed to be used consistently as a rule, guideline, or definition. They help to make life simpler and increase the reliability and the effectiveness of many of the goods and services we use" [35] . Some economists have given various definition of standards, among them are; Alfred Marshall who defined standards as "customs" that recorded technical progress, firstly in an informal manner and then in a formal manner starting in industrial era [36] . David et al. (1990) defined standards as "a set of technical specifications adhered to be a producer, either tacitly or as a result of formal agreement" [37] . Richard et al. (1995) defined standards as "agreed external points of reference to which the physical and performance characteristics of technologies can be compared" [38] .

From the above literatures we can understand that standards are published documents that establish specifications and procedures designed to maximize the reliability of the materials, products, methods, and/or services people use every day. Standards address a range of issues, including but not limited to various protocols to help maximize product functionality and compatibility, facilitate interoperability and support consumer safety and public health.

Standards form the fundamental building blocks for product development by establishing consistent protocols that can be universally understood and adopted. This helps fuel compatibility and interoperability and simplifies product development, and speeds time-to-market. Standards also make it easier to understand and compare competing products. As standards are globally adopted and applied in many markets, they also fuel international trade.

It is only through the use of standards that the requirements of interconnectivity and interoperability can be assured. It is only through the application of standards that the credibility of new products and new markets can be verified. In summary standards fuel the development and implementation of technologies that influence and transform the way we live, work and communicate.

Process of Standards Development

The process of developing a standard is typically facilitated by a Standards Development Organization (SDO), which adheres to fair and equitable processes that ensure the highest quality outputs and reinforce the market relevance of standards. SDOs, such as IEEE, IEC, ISO, and others, offer time-tested platforms, rules, governance, methodologies and even facilitation services that objectively address the standards development lifecycle, and help facilitate the development, distribution and maintenance of standards. While the goals of each SDO are essentially the same, each SDO applies its own rules, processes, and terminology to the standards development process. Typically, each SDO is comprised of boards, committees and staff who establish and maintain the policies, procedures and guidelines that help ensure the integrity of the standards development process, and the standards that are generated as an outcome of this process.

The development of a new standard is typically triggered by a formal request, submitted to an SDO by a Sponsoring Body (individual or entity, such as an industry society) for review and evaluation. The SDO mandates, oversees, and helps facilitate the process for standards development. The Sponsor for the standards project assumes responsibility for the respective area of standards development, including the organization of the standards development team and its activities. Once the SDO approves the request to develop a new standards development project, the Sponsor follows the SDOs rules and processes to recruit and assemble a collaborative team or "Working Group" to engage in active standards development (note: the term "Working Group" is an IEEE term. Working Groups may be called different names by the various SDOs, and may follow slightly different processes). Working Groups are comprised of individuals and/or entities (people, companies, organizations, non-profits, government agencies) who volunteer to support the development of standards [39] .

Collectively, these volunteer participants carry a specific interest in a specific area of development as producers, sellers, buyers, users and/

or regulators of a particular material, product, process or service. When a Working Group is formed the Working Group officers may either be elected by the Working Group members or appointed by the Sponsor. Consult the Sponsor's policies and procedures and the Sponsor's Working Group policies and procedures for details. Working Group officers oversee the standards development project in adherence to SDO rules and process, and remain accountable to the project Sponsor and the governance structure of the SDO itself.

Based on the rules and criteria established by the SDO, participants may contribute at varying levels to the standards development process. For example, the IEEE Standards Association (IEEE-SA) has established rules related to membership and participation, and employs a separate "Entity Standards Development Process" for standards that are sponsored by entities (such as corporations, governments, non-profits, associations or other organizations, etc.). Such rules help ensure that highly dedicated individuals lead participation and no one interest dominates the standards development process. Working Groups leverage these rules and guidelines and establish their own individual, organizational, communications and meeting structures, and govern work process, activities, consensus building, decision making, balloting and even financial reporting in accordance with SDO rules. To build consensus through democratic means, participants engage in meetings, draft and review position pieces, create and review presentations, examine data and engage in active discussion and debate to resolve outstanding issues.

These activities fuel the gradual definition of each standard, which is compiled into a draft standard that may undergo multiple revisions. Once a draft standard has been finalized, reviewed, and approved by the Working Group, it is submitted to the Sponsor for Sponsor balloting. Upon successful completion of the Sponsor ballot, the draft is submitted to the Review Committee (RevCom). The balloted draft is reviewed by RevCom and then submitted to the Standards Board for approval. After submission, review and acceptance, the approved standard is published and made available for distribution and purchasing within in a number of outlets, including through the SDO itself [39] [40] .

It is important to remember that standards are "living documents", which may initially be published and iteratively modified, corrected, adjusted and/or updated based on market conditions and other factors. At any given point in time, therefore, a standard may be referred to as having a number of different "status" classifications.

These include (see Figure 4):

- Approved Project—An initial project request is approved, in stages of group formation.
- Active Project—An active standards development project.
- Withdrawn Project—A cancelled standards development project.
- Approved Standard—The standard is approved and published for public use.
- Withdrawn Standard—The standard is no longer market relevant or active.
- Superseded Standard—The standard has been replaced by a new standard.

In keeping with the standards development lifecycle, Working Groups may also go through periodic stages of activity or dormancy. Depending on where a standard is in its lifecycle a standard may be accompanied by supplemental documents that are produced by its respective Working Group. These may include errata (which address errors in the standard), amendments (which modify sections of the standard), corrigenda (which only correct errors or ambiguities in a standard), handbooks, tutorials and other related materials. Supplemental documents help interested parties better understand and apply the standard [39] .

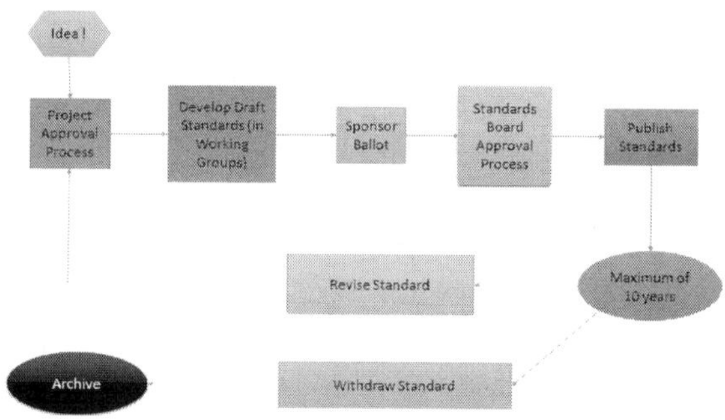

Figure 4. Process of standards development.

Roles and Functions of Standards

Standards are established through government policies or consumer protection organizations to protect the consumers of various goods and services as most consumers have little or no choice in what they are offered. In rural communities in developing countries, consumers do not generally have the luxury of comparing features and selecting their suppliers or products from the Internet. Hence, the role of standards is to ensure that whatever product or service is provided to the consumers is fit for purpose, safe and has value. An important aspect of this protection is to ensure the product or service delivers as claimed, performs as specified, and is reliable, durable and safe. In many cases renewable energy standards and conformity assessment can be the catalyst in providing alternatives to systems which are sometimes unsafe, operate with low efficiency, and use fossil fuel energy with possible detrimental effects on health. This is achieved by providing confidence and trust in renewable energy products and those who provide energy-related services. Standards also have the ability to allow those not typically trained in these energy sources to reach a level of understanding that allows them to provide, install or operate systems for themselves or under reduced supervision. This is achieved by providing guidance and best practice in designing, specifying, installing and maintaining the systems. A good example is the set of international standards for rural electrification. Standards also provide an effective framework for harmonizing information flow, understanding technical product design, manufacturing and service requirements, as well as establishing common rules and requirements. Standards should enable all these functions to take place while ensuring there is flexibility for the product, service, system provider and user [35] .

The "De Facto" and "De Jure" Standards

A "de facto" standard is one created almost by default because of one dominant player in that industry. A "de facto" standard can also be described as a standard that achieved its status as a result of market competition. For example, Windows for personal computers or VHS for video cartridge recorders are typical de facto standards. A process that a product becomes a standard after it is brought to the market and exposed to competition is called the "classical model". For this instance, such case as Microsoft which maintains monopoly in the market is quite unusual. It is more common that several companies form a group, and those several groups make competition among each other and the winner is determined. A process that a winner is determined after products are brought to the market imposes a heavy burden on companies as they already have completed their capital investment. Consumers also get lost, as even if they purchased products, those styles may no longer be seen in the market some time later. A good example is that consumers felt inconvenience if they had bought video tapes of beta system for some time as they only saw those of VHS system some time later and those tapes previously bought could not be used anymore. Other de facto standards include QWERTY keyboard, Microsoft Word format for documents and Adobe Postscript languages for laser printers [41] [42] .

The opposite of an informal "de facto" standard is the "de jure" standard, which means "by force of law". The de jure standards are standards that are developed after going through a certain procedure or official organization. Those standards developed by ISO or IEC are developed through a procedure that experts gather from all over the world and discuss on standardization proposals, which are finally approved through voting by national standards institutes. Japan's standards JIS are also developed going through a procedure that JISC first deliberate a proposal and competent ministers finally make decision to incorporate the proposed standards in the JIS. Korea's Standards KS are developed by Korean Standards Association which is a public organization under the Korean Ministry of Trade, Industry & Energy (MOTIE) to carry out industrial standardization, certification, and quality promotions thereby contributing to national economy of Korea. However, de jure standards are developed through a transparent and democratic procedure. As companies bring their products to the market based on de jure standards, they do not impose excessive burden on corporations or consumers. However, development of de jure standards takes a long time. They are not suitable for products with short cycle such as IT-related products. Examples of de jure standards include Internet TCP/IP protocol, Wireless 801.11n and Unicode international character encoding [41] [42] .

As a result, a new type of standards "forum standards" has been introduced recently, which adopts the positive parts of both de fact standards and de jure standards. Influential companies get together to form a standard before bringing products to the market, so that neither companies nor consumers need to take risks [41] .

Standards Regulatory Agency in Nigeria

The only standards regulatory agency in Nigeria is the Standards Organization of Nigeria (SON) which was established by Act. No. 56 of 1971 which vested it with the authority for: standards elaboration, specification, quality assurance system commodities, manufactured industrial and imported products and services. The Act No. 20 of 1976 which amended the previous one conferred on the Honorable Minister of Industry the power to declare Mandatory Industrial Standards in respect of products or processes recommended by the Nigerian Standards Council.

The Act No. 32 of 1984 changed the name of the Organization to Standards Organization of Nigeria (SON) from Nigeria Standards Organization (NSO) to eliminate conflicting identity with the then Nigerian Security Organization. Finally, the Act No. 18 of 1990 conferred on SON partial autonomy from the Ministry of Industry. This amendment gave far-reaching transformation to the organization succession and a common seal, and may sue or be sued in its corporate name. The statutory functions of Standards Organization of Nigeria by Section 3, subsections (1) of 1971 Act No. 56 are as follows:

- To organize test and do everything necessary to ensure compliance with standards designated and approved by the council;
- To undertake investigations as necessary into the quality of facilities, materials and products in Nigeria, and establish a quality assurance system including certification of factories, products and laboratories;
- To ensure reference standards for calibration and verification of measures and measuring instrument;
- To compile an inventory of products requiring standardization;
- To compile Nigeria Industrial Standards;
- To foster interest in the recommendation and maintenance of acceptable standards by industry and the general public;
- To develop method for testing of materials, supplies and equipment including items purchased for use of departments Government of the Federation or State and Private establishment;
- Register and regulate standard marks and specification;
- To undertake preparation and distribution of standard samples;
- To establish and maintain such number of laboratories or other institutions as may be necessary for the performance of its functions under this Act;
- To compile and publish general scientific or other data:
- Resulting from performance of its function under this Act; or
- From other sources when such data are of importance to scientific or manufacturing interest or to the general public and are not available;
- To advise departments of government of the Federation or state on specific problems relating to standards.

- Nigeria represented by SON is a member of following International Standards Bodies:
- International Organization for Standardization (ISO);
- International Electrochemical Commission (IEC);
- African Organization for Standardization (ARSO).

Presently, Nigeria (SON) is a member of the International Organization for Standardization (ISO) with Participating (P-Member) and Observer Member (O-Member) in various Technical Committees (TC) and Project Development Committees (PDCs) [43] . Among all the standards available in the SON, there is none that is responsible for the standards of renewable energy technologies as at present. However, the next section describes a developmental process in which a renewable energy standard could be formulated and adapted by the SON.

DEVELOPMENT OF RENEWABLE ENERGY STANDARDS

Currently, the SON maintains 33 Standing Technical Committees (TC) which also have sub technical committees that deals with specific issues. The various TC usually creates a forum wherein the various experts are harnessed in the development of standards. A draft standard is usually prepared by the SON Technical Officers in the Organization depending upon the priority or work programme of the directorate and on the need or demand from the stakeholders. The officers, thereafter collect the relevant scientific data related to the draft standard under development from technical literatures including journals, laboratory tests results and factory inspections among others. The draft is then sent out to the TC members for review. The comments are received and collate the and the meeting of the relevant TC is held to review the comments and evaluated all quality parameters being prescribed are accepted by consensus to be adequate for approval by the Nigeria Standards Council or in its absence the Honorable Minister of Supervision Ministry of the Standards Organization of Nigeria, Approval of National Standards. The responsibility of authorization of National Standards as per the Act establishing SON is vested in the Organization's Governing Council also referred to as Nigeria Standards Council. In absence of the Council, the Honorable Minister, Federal Ministry of Trade and Investments has the responsibility to approve a standard for use. Apart from development of new standards, existing standards are revised or reviewed so that the outdated matches with the current quality parameters. The procedure for review of standards is the same with that of drafting new ones. Standards are reviewed after three years, if there is technological development in the sector concerned or where inadequacies are observed. When standards reviewed and are no longer relevant for the purposes they were developed are withdrawn [43] .

As earlier discussed in this study, the SON does not have any form of standards for renewable energy technologies present. However, development of renewable energy standards would be a welcomed development. The testing and proper certification of renewable energy equipment for off-grid or small scale application will still face some challenges. These challenges could come from the political class or importers who import substandard products into the country. Most renewable energy products are imported from China into the Nigerian market without proper product certification. Some of the imported technologies include solar panels, inverters, energy storage systems, small hydropower (SHP) units and wind turbines.

The deployment of renewable energy depends on some issues which are external to standards like Feed- in-Tariffs (FiTs), planning, environmental factors, building regulation and economic development. Standards will be used along with various policies, certification and regulation to open up renewable energy market.

Developing renewable energy standards involves high cost and is time consuming, this has hindered actors from participating in the process of renewable energy standards development. Development of renewable energy standards can be achieved if the SON collaborate with the ECN and other relevant bodies in engaging more external stakeholders who will make use of reports, case studies, research articles and various workshops to ensure that proper standards are developed for renewable energy technologies. These activities will improve the mix of experts that will in turn improve the quality of the standards under development. The SON will need to increase its TC to accommodate experts in the field of renewable energy. The SON should also engage specific projects related to renewable energy standards development as well as promote knowledge dissemination in renewable energy technologies.

Cost of Standards

Some challenges are faced by standard bodies in the developing countries is have the right standards in the first place and to have access to purchase the standards and also to keep them up to date. The cost implications for the standard organizations in the developing countries and emerging markets can make the difference between adopting the already existing internationally developed standards, producing local requirements or the complete rejection of the standards. The cost of standards may be seen as a barrier to the adoption of standards, but in the case of Nigeria, it should be seen as means to ensure that her citizens are protected from poor quality renewable energy technologies manufactured with the best practice of standardization. However, some recommendations for small renewable energy and hybrid systems are provided in Table 2 for rural electrification (see [35]).

Recommended Standards for Renewable Energy Technologies

Below are some recommended standards currently available through the International Organization for Standardization (ISO). These recommended standards should be used as an immediate measure to address the issue of substandard importation of renewable energy technology.

- ISO/TC 180 Solar Energy

Standardization in the field of solar energy utilization in space and water heating, cooling, industrial process heating and air conditioning.

- IEC/TC 82 Solar Photovoltaic Energy System

Includes the entire field from light input to a photovoltaic cell to and including the interface with the electrical system(s) to which energy is supplied.

- IEC/TC 88 Wind Turbines

These standards address design requirements, engineering integrity, measurement techniques and test procedures. Their purpose is to provide a basis for design, quality assurance and certification. The standards are concerned with all subsystems of wind turbines, such as mechanical and internal electrical systems, support struc-

Table 2. Recommendations for small renewable energy and hybrid systems for rural electrification [35] .

Standards No.	Standard Name	Cost of Standards (USD), [Cost in NGN]*
IEC/TS 62257-9-1	Recommendations for small renewable energy and hybrid systems for rural electrification—Part 9-1: Micro-power systems	252.31 [41441.58]
IEC 60364 (all parts)	Low-voltage electrical installations Note: All parts would equate to 32 standards documents	4284.00 [703641.27]
IEC 60364-5-53:2001 included in all parts above	Electrical installations of buildings—Part 5-53: Selection and erection of electrical equipment—Isolation, switching and control	Not Available**
IEC 60529	Degrees of protection provided by enclosures (IP Code)	262.50 [43115.27]
IEC/TS 62257-2:2004	Recommendations for small renewable energy and hybrid systems for rural electrification—Part 2: From requirements to a range of electrification systems	283.22 [46518.51]
IEC/TS 62257-4:2005	Recommendations for small renewable energy and hybrid systems for rural electrification—Part 4: System selection and design	188.81 [31011.79]
IEC/TS 62257-5:2005	Recommendations for small renewable energy and hybrid systems for rural electrification—Part 5: Protection against electrical hazards	188.81 [31011.79]
IEC/TS 62257-6:2005	Recommendations for small renewable energy and hybrid systems for rural electrification—Part 6: Acceptance, operation, maintenance and replacement	94.41 [15506.72]
IEC/TS 62257-7-1:2006	Recommendations for small renewable energy and hybrid systems for rural electrification—Part 7-1: Generators—Photovoltaic arrays	293.71 [48241.47]
IEC/TS 62257-7-3:2008	Recommendations for small renewable energy and hybrid systems for rural electrification—Part 7-3: Generator set—Selection of generator sets for rural electrification systems	209.79 [34457.73]
IEC/TS 62257-9-2:2006	Recommendations for small renewable energy and hybrid systems for rural electrification—Part 9-2: Micro-grids	251.75 [41349.6]

IEC/TS 62257-9-4:2006	Recommendations for small renewable energy and hybrid systems for rural electrification—Part 9-4: Integrated system—User installation	25.87 [4249.11]
Total		6435.18 [1056969.7]

Note: *NGN means Nigerian naira. The currency conversation was made on 4th October, 2014; **The cost information was not available at the time of publication. tures and control and protection systems. They are intended to be used together with appropriate IEC/ISO standards.

- IEC/TC 4

Hydraulic rotating machinery and associated equipment allied with hydro-power development.

- ISO/TC 238 Solid Biofuels

Standardization of terminology, specifications and classes, quality assurance, sampling and sample preparation and test methods in the field of raw and processed materials originating from arboriculture, agriculture, aquaculture, horticulture and forestry to be used as a source for solid biofuels. Excluded: areas covered by ISO/TC 28/SC 7 Liquid biofuels, ISO/TC 193 Natural gas.

- ISO/PC 248 Sustainability Criteria for Bioenergy

Standardization in the field of sustainability criteria for production, supply chain and application of bioenergy. This includes terminology and aspects related to the sustainability (e.g. environmental, social and economic) of bioenergy.

- ISO/TC 255 Biogas

Standardization in the field of biogas.

Recommendations to Standards Organization of Nigeria

The Standards Organization of Nigeria which is the standard body in Nigeria will need to make some adjustment to its policy in order to protect the citizens of Nigeria from the influx of poor quality renewable energy products into the country. Some recommendations are given below:

- Develop standard compliance cooperation with foreign countries where most of the renewable energy products are manufactured;
- Development of renewable energy standards with the cooperation of international standard bodies in the field of renewable energy;
- Promote the training of experts through development programmes,

constructing standard databases, designating a permanent agency or commission for the promotion and implementation of renewable energy product standard;

- Implement laws for the promotion of renewable energy standardization plans, organization, budget and human resource development;
- Revise a number of laws and articles to ensure that renewable energy standardization is embedded in the political framework;
- Ensure that each renewable energy product imported into the country by an importer are given After Sales Service (ASS) for up-to 5 years from the time of purchase to the consumer.

CONCLUSIONS

For sustainable energy to be achieved in any country, clean energy development has to be a national priority. The issues of climate change have alerted most governments around the world to mitigate global warming by moving towards renewable energy adoption and incorporation into their energy mix. The manufacturing of renewable energy technologies like solar panels, solar thermal, inverters, energy saving systems, various kinds of wind turbines and hydropower systems has increased in recent times. Most countries like South Korea, Japan, China and the US manufacture these renewable energy technologies according to their own specification and standards and export it to other countries that do not have the technical expertise or resources to produce theirs. While this may sound good, some countries do manufacture substandard renewable energy technologies and export it to developing countries that may not have proper standards bodies to check the imported products into their market.

Nigeria has abundant renewable energy resources but do import the technology which is able to harness the renewable energy potential. These technologies are usually substandard and are dangerous to her citizens. This study has discussed the necessity of the development of standards for renewable energy technology which is imported, stressed the need to develop local content in the field of renewable energy and given some recommended standards that could be applied. Nigeria has a standard body which is the Standards Organization of Nigeria (SON) but no standards have been made available for renewable energy technologies. However, the country depends on standards set by other countries that export the technologies.

However, some immediate measures should be put in place by the SON to protect Nigerian citizens from the various substandard renewable energy technologies which include solar panels, inverters, and energy-saving systems. As an immediate measure, the SON should ensure that the technologies which are imported into the country are checked for international standards which are usually granted by International Electrochemical Commission (IEC) and also apply the given recommendation in this study. This will stop the influx of substandard renewable technologies while the SON develops its own standards.

Vincent Nnaemeka Emodi, Samson D. Yusuf, Kyun-Jin Boo This study strongly believes that the development of standards for renewable energy technologies is of great importance to secure the citizens of Nigeria and also to fast track the development of renewable energy in Nigeria. The study focused on the necessity for the development of standards for renewable energy technologies and recommended some standards but did not propose any standards. Thus, more studies should be undertaken to come up with a proper standard that can be submitted to the SON.

REFERENCES

1 Sambo, A.S. (2005) Renewable Energy for Rural Development: The Nigerian Perspective. ISESCO Science and Technology Vision, 1, 12-22.

2. Oyedepo, S.O. (2012) Energy Efficiency and Conservation Measures: Tools for Sustainable Energy Development in Nigeria. International Journal of Energy Engineering, 2, 86-98.

3. Okafor, E.N.C. and Joe-Uzuegbu, C.K.A. (2010) Challenges to Development of Renewable Energy for Electric Power Sector in Nigeria. International Journal of Academic Research, 2, 211.

4. Patlitzianas, K., Flamos, A., Doukas, H., Kagiannas, A. and Psarras, J. (2007) Promoting Renewable Energies and Energy Efficiency through the CDM Funding Options. 2004 New and Renewable Energy Technology, 1.

5. European Commission (2003) Second ECCP Progress Report: Can We Meet Our Kyoto Targets? European Commission, Brussels.

6. Why Solar Energy Is Underutilized in Nigeria—Suleiman (2014). www.leadership.ng/news/347736/solar-energy-underutilised-nigeria-suleiman

7. Shaaban, M. and Petinrin, J.O. (2014) Renewable Energy Potentials in Nigeria: Meeting Rural Energy Needs. Renewable and Sustainable Energy Reviews, 29, 72-84. http://dx.doi.org/10.1016/j.rser.2013.08.078

8. Vincent, E.N. and Yusuf, S.D. (2014) Integrating Renewable Energy and Smart Grid Technology into the Nigerian Electricity Grid System. Smart Grid and Renewable Energy, 5, 220-238. http://dx.doi.org/10.4236/sgre.2014.59021

9. Sambo, A.S. (2009) Strategic Developments in Renewable Energy in Nigeria. International Association for Energy Economics, 16.

10. Uduma, K. and Arciszewski, T. (2010) Sustainable Energy Development: The Key to a Stable Nigeria. Sustainability, 2, 1558-1570. http://dx.doi.org/10.3390/su2061558

11. Nadabo, S.L. (2010) Renewable Energy as a Solution to Nigerian Energy Crisis.

12. Energy Commission of Nigeria (ECN) (2009) Renewable Energy Master Plan.

13. Ilenikhena, P.A. and Ezemonye, L.I.N. (2010) Solar Energy Applications in Nigeria. WEC MONTREAL.
14. Yearly Average of Daily Sun in Nigeria. http://www.soda-is.com/eng/nigeria_en.html
15. Adeoti, O., Oyewole, B.A. and Adegboyega, T.D. (2001) Solar Photovoltaic-Based Home Electrification System for Rural Development in Nigeria: Domestic Load Assessment. Renewable Energy, 24, 155-161. http://dx.doi.org/10.1016/S0960-1481(00)00188-9
16. Bugaje, I.M. (2006) Renewable Energy for Sustainable Development in Africa: A Review. Renewable and Sustainable Energy Reviews, 10, 603-612. http://dx.doi.org/10.1016/j.rser.2004.11.002
17. Oyedepo, S.O. (2012) On Energy for Sustainable Development in Nigeria. Renewable and Sustainable Energy Reviews, 16, 2583-2598. http://dx.doi.org/10.1016/j.rser.2012.02.010
18. Energy Commission of Nigeria (2006) Energy Commission of Nigeria (ECN) Renewable Energy Master Plan. Energy Commission of Nigeria, Government of Nigeria, Abuja.
19. Nigeria Wind Energy. http://www.neenigeria.com/Nigeria_wind_NEW.png
20. Bada, H.A. (2011) Managing the Diffusion and Adoption of Renewable Energy Technologies in Nigeria. World Renewable Energy Congress, Linkoping, 8-13 May 2011, 2642-2649.
21. Mohammed, Y.S., Mustafa, M.W., Bashir, N. and Mokhtar, A.S. (2013) Renewable Energy Resources for Distributed Power Generation in Nigeria: A Review of the Potential. Renewable and Sustainable Energy Reviews, 22, 257-268. http://dx.doi.org/10.1016/j.rser.2013.01.020
22. Sambo, A.S. (2009) The Place of Renewable Energy in Nigeria Energy Sector. World Future Council Workshop on Renewable Energy Policies, Addis Ababa, 10 October 2009, 5.
23. Ohunakin, O.S. (2010) Energy Utilization and Renewable Energy Sources in Nigeria. Journal of Engineering and Applied Sciences, 5, 171-177. http://dx.doi.org/10.3923/jeasci.2010.171.177
24. The Nigerian Water Ways (2014). http://www.safty4sea.com/

25. Iloeje, O. (2002) Renewable Energy Development in Nigeria: Status & Prospects. Proceedings of a National Workshop on Energizing Rural Transformation in Nigeria: Scaling up Electricity Access and Renewable Energy.
26. Ikuponisi, F.S. (2004) Status of Renewable Energy in Nigeria. One Sky Energetic Solutions Conference, Nigeria, 21-27 November 2004, Background Paper.
27. Abumere, S.I., Okafor, S.I. and Oluwasola, O. (2002) Rural Infrastructure and the Development Process in Rural Nigeria (No. 36). Development Policy Centre.
28. Sambo, A.S. (2010) Renewable Energy Development in Nigeria: A Situation Report. World Future Council/Strategy Workshop on Renewable Energy, Accra, 21-24 June 2010.
29. Spalding-Fecher, R., Winkler, H. and Mwakasonda, S. (2005) Energy and the World Summit on Sustainable Development: What Next? Energy Policy, 33, 99-112. http://dx.doi.org/10.1016/S0301-4215(03)00203-9
30. Efurumibe, E.L. (2013) Barriers to the Development of Renewable Energy in Nigeria. Scholarly Journal of Biotechno- logy, 2, 11-13.
31. Standardization (2014). http://www.encyclopedia2.thefreedictionary.com/
32. Sykes, A.O. (1995) Product Standards for Internationally Integrated Goods Markets. Brookings Institution, Washington DC.
33. International Standard Organization (ISO). http://www.iso.org/
34. World Trade Organization (WTO). http://www.wto.org/
35. REP, IRENA (2013) International Standardisation in the Field of Renewable Energy.
36. Marshall, A. (1919) Industry and Trade: A Study of Industrial Technique and Business Organization and of Their Influences on the Condition of Various Classes and Nations. MacMillian and Co. Limited, London, New York.
37. David, P.A. and Greenstein, S. (1990) The Economics of Compatibility Standards: An Introduction to Recent Research. Economics of Innovation and New Technology, 1, 13-41.
38. Hawkins, R., Mansel, R. and Skea, J., Eds. (1995) Standards, Innovation and Competitiveness: The Politics and Economics of Standards in Natural and Technical Environment Edward Elgar, Aldershot, Brookfield, p. 1.

39. IEEE Standards Association (2014) How Are Standards Made? http://www.standards.ieee.org/develop/process.html
40. Try Standards: Global Standards Education and Standards Search. http://www.trystandards.org/mod/page/view
41. The Difference between De Facto Standards and De Jure Standards (2014). http://www.ncos.gr.jp/English.files/e_newpage51
42. De facto and De Jure Standards (2014). http://www.teach-ict.com/as_a2_ict_new/ocr/A2_G063/333_networks_coms/standards/miniweb/pg3
43. Standards Organization of Nigeria (2014). http://www.son.gov.ng/

16

The Development of the Renewable Energy Technologies in Spain

Félix Hernández[1], Miguel Hernández-Campos[2]

[1]Institute of Economics, Geography and Demography (Spanish National Research Council), Madrid, Spain;
[2]Foundation Gómez Pardo, Madrid, Spain.

ABSTRACT

This article examines the development of the four main renewable energy technologies (RET) in Spain in the latest years: biomass, small hydro (SH), solar photovoltaic (solar PV) and wind. It has been studied the variation of the working time per MW installed available from the on line data base of Spanish National Commission on Energy (SNCE), in the national context and in each of the Autonomic Communities (AACC). We have also obtained the marginal cost curve of the global RE. The main conclusions of the study are that Spain is effective meeting the RE generation target but not efficient in costs and in short term it is not any RETs that can achieve competitive price comparing to the electrical market.

Citation: F. Hernández and M. Hernández-Campos, "The Development of the Renewable Energy Technologies in Spain," Smart Grid and Renewable Energy, Vol. 2 No. 2, 2011, pp. 110-115. doi. 10.4236/sgre.2011.22013.

Keywords: Renewable Energy Technologies, Working Time per MW, Feed-in Tariff, Marginal Cost Curve of the Global RE

INTRODUCTION

Most of EU governments have adopted measures aimed in promoting of the Renewable Energy Technologies (RETs). The degree of success of these measures has been variable in terms of efficiency in costs and deployment effectiveness [1]. Spain has been cited as an example for its success to get more RETs in place through a feed-in tariff (FIT) system, but there are serious concerns about their rising costs. In fact, FIT are the most effective system to a rapid deployment of RETs, but when it is in place the installations capacity (in MW) increase spectacularly, often with undesirable effects. For example, in the future some RETs could not survive financially unless their subsidies come at the expense of customers. Wang [2] already warned that Spain had to drastically reduce its FIT payments to Solar PV projects and impose caps on annual installed capacity of this TER.

It is necessary therefore to find a comparative parameter that gives some light about the rhythm installed capacity in relation with the number of annual full load hours of the RETs. Couture and Gagnon [3] speak about the per-kWh payments adjusted on the number of annual full load hours, referring this term to the hypothetical number of hours in which a RET would need to operate to produce its annual production.

The current paper studies the development of the different RETs in Spain in the latest years. It embraces the following special regimen[1] technologies: biomass, small hydro (SH), solar photovoltaic (solar PV) and wind. All of these are subsidized by the administration through a FIT system, added to the cogeneration, wastes and wastes treatment[2], not include in this paper.

Based on the empirical data provided from www. cne.es [4] by the Spanish National Commission on Energy (SNCE) reported for March 10th, 2010, all the RETs have been analyzed developing an accurate tool that correlates the variation of working hours recorded during the annual production of GWh per MW installed, according to the Spanish and all Autonomic Communities (AACC). The evolution of the equivalent FIT[3] will be examine as well as a percentage of the total retribution by the sold energy and the contribution per MW for all the RETs. Finally, a marginal cost curve has been obtained from the global Renewable Energy in Spain.

VARIATION OF WORKING HOURS PER MW INSTALLED

As a comparative parameter of the developing of all the different RETs it has been chosen the calculated working hours according to the sold GWh per MW installed, from the on line data base of SNCE. Their development with time makes possible to estimate the fulfilment of each RETs in the electric market. In fact the working hours of RETs represents the amortized income with the sale of the produced work in GWh by the whole power plants of each RETs; the installed capacity in MW[4] is the total capital to be amortized; its quotient means the amortized index in hours/MW units. The increase of this index with time belongs to a positive integration of the RET in the electric market system. Oppositely, a decrease means lack of competitiveness of the RET in the energy market.

National Context (National Market)

The **Figure 1**[5], represents the fluctuations between working hours recorded in the sale of RET per MW installed, in the period 2004 - 2009, of the RETs mentioned above: biomass, SH, solar PV and wind.

As additional information from the development of the working hours per MW of the sold RETs, the following **Table 1** represents the maximum variation in percentage from the main value for all the RETs.

From the brief analysis of the **Figure 1** it can infer the following items:

- The variable working hours per MW tends to change for each RET:
- Almost draws a symmetric function in the SH[6].
- Describes a decreasing smooth line for biomass and wind and a decreasing sharply slope in the Solar PV, due to the following causes (alternative or simultaneous):

Lack of planning between the development of power plants (installed capacity increasing) and the sold working hours per MW covered by the FIT system (decreasing).[7]

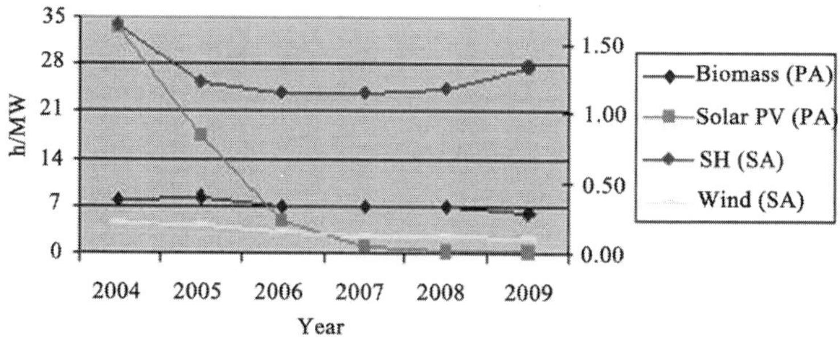

Figure 1. Working hours per MW.

Table 1. Maximum variation of working hours per MW respect the average.

RET	Variation respect the average (%)
Biomass	19.5
SH	36.8
Wind	60.5
Solar PV	346.8

- Gradual reduction in the technological potential in the RETs.
- Problems to adjust the energy generation and the electrical demand.
- Decrease in the electrical demand.
- Punctual fall in the renewable resources.

The **Table 1** shows that in the case of the Solar PV, the working time variation respect to the average is large (346.8), which means sharply variations in the energy generated for this technology in the period 2004 - 2009, that is to say, this represents an unsustainable developing of this RET [6] linked to its high FIT[8].

AACC Context

The **Table 2** includes the variation η between the sold working hours per MW installed in the period 2007 - 2009[9] (WH_{2007} - WH_{2009}), by the SNCE, with regard the first year recorded 2007 (WH_{2007}). The variation has been calculated for all the RETs and AACC from the on line data base of SNCE.

From the **Table 2** we have built the **Figure 2**. It shows the variation of η, divided in regular intervals, in the period 2007 - 2009 for all AACC. The Figure illustrates the change in relative frequency of the RETs, number of times the value η appears in the corresponding row, in relation to the interval of hours per MW considered in the Table. In fact, each RET has a maximum for a different interval: 56.3% for SH and $\eta < 0$; 40.0% for biomass and 46.7% for wind and $25\% \leq \eta \leq 50\%$, and 41.2% for Solar PV and $50\ \% \leq \eta \leq 75\%$.

Table 2. Percentage variation of η (2007 - 2009).

La Rioja	SH		22.3			
	Solar PV					79.5
	Wind			30.9		
	Biomass			34.3		
Madrid	SH		10.9			
	Solar PV				60.8	
	Biomass					81.3
Murcia	SH		14.2			
	Solar PV			32.4		
	Wind			33.1		
	Biomass	-23.0				
Navarra	SH			36.7		
	Solar PV			48.6		
	Wind		24.1			
	Biomass			35.5		
Pais Vasco	SH	-9.4				
	Solar PV			43.3		
	Wind			29.0		
	Biomass			35.8		
Valencia	SH	-271.5				
	Solar PV			37.4		
	Wind				61.7	
Cantabria	SH		13.1			
	Solar PV				52.4	
	Wind	-4.5				
	Biomass		15.1			
Castilla La Mancha	SH	-9.4				
	Solar PV				54.3	
	Wind		22.7			
	Biomass	99.5				
Castilla León	SH	-0.1				
	Solar PV				58.1	
	Wind			43.4		
	Biomass		13.7			
Cataluña	SH	-66.9				
	Solar PV				60.8	
	Wind				51.8	
Extremadura	SH	-20.8				
	Solar PV	326.1				
	Biomass		22.8			
Galicia	SH	-17.5				
	Solar PV				71.0	
	Wind			30.0		
	Biomass			44.2		

AACC	RET	Percentage variation $\eta = (WH_{2007} - WH_{2009})/WH_{2007}*100$				
		$\eta < 0$	Interval variation of η%			
			0 - 25	25 - 50	50 - 75	75 - 100
Andalucía	Biomass			35.8		
	SH	-44.7				
	Solar PV					86.5
	Wind			45.9		
Aragón	Biomass				51.1	
	SH	-5.7				
	Solar PV					89.6
	Wind			25.1		
Asturias	Biomass				52.7	
	SH		14.2			
	Solar PV				59.0	
	Wind				52.8	
Baleares	Solar PV					97.3
	Wind				58.8	
Canarias	Biomass	-105.3				
	SH				68.0	
	Solar PV	-234.0				
	Wind		16.1			
	Biomass			32.4		

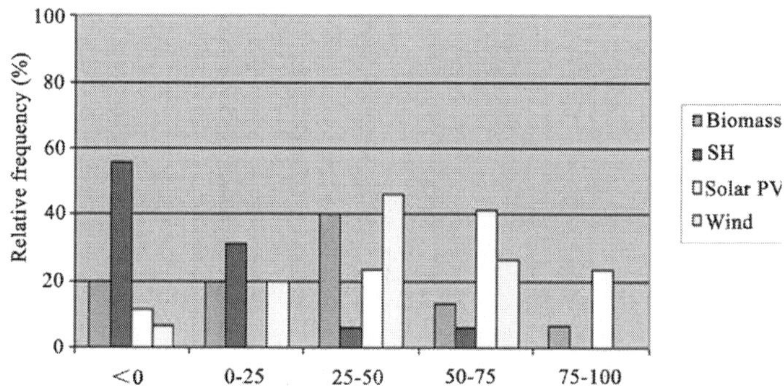

Figure 2. Percentege de variation hours per MW (2007 - 2009).

The **Table 2** and the **Figure 2** show the state of development of RETs in all AACC in relation to η. The negative sign points out that the degree of development it trends to rise, and decreasing as η, variation between the sold working hours per MW installed in the period 2007 - 2009, is growing up. Thus, we can infer that SH is the technology which development is raising the most among all the RETs and Solar PV is decreasing in the majority of AACC. We can also establish the following particularities from the η values:

In Canarias[10], Castilla León and Navarra the biomass has grown 105.3%, 99.5% and 23.0% respectively.

- In Canarias the SH has decreased 68.0%.
- Exceptionally, the Solar PV has grown in Canarias (234.0%) and Extremadura (326.1%), and the wind in Cantabria (4.5%).
- In Navarra, sold working hours of SH, Solar PV and wind per

MW has decreased harmoniously, becoming a rational initiative for the legal authorization granting a regulated utility to construct a RETs power plant In Canarias the SH has decreased 68.0% (see note 6).

EVOLUTION OF THE NET EQUIVALENT FIT PER MW INSTALLED

The concept of feed in tariff it is really useful to check out the development of each RET comparing their competitiveness to the whole electricity market. According with definition above established by the SNCE, the equivalent feed in tariff is the difference between the total retribution received for each RET and the total energy multiplied by the annual regular price in the electrical market. Therefore it is the cost of RET calculated from the on line data base of SNCE by reference of the conventional energy incorporating the effect of the support policies in a straightforward way. Costs are expressed in terms of the minimal price that the investor has to obtain from the market over the lifetime of the production capacity in order to make the construction of additional capacity (or the production with existing capacity) attractive.

In order to calculate the equivalent FIT it has just been considered the power plants that had sold the energy directly through the electrical market or by a private agreement with a distribution company with negotiated fares and other alternative markets. So far, the net equivalent FIT it has been obtained as the subtraction of the equivalent FIT minus the avoided emissions[11] cost for all the RETs[12] analyzed.

The temporally development of the net equivalent FIT per MW installed, it works out as a very useful tool for checking how alike it is each RET from the competitive electricity market. This is considerate an effective indicator to analyze the prospects for diffusion of each RET, equivalent to a learning curve [7]. The **Table 3" targct="_self"> Table 3** shows the net equivalent FIT per MW installed during the period 2004 - 2009 that belongs to the Spanish situation.

From **Table 3** it can be achieved two main conclusions:

- The development of the FIT per MW installed shows maximums in the latest year of 2009 and a minimum value en the first collected years 2004 and 2005.
- Therefore, it is not achieved any competitive price comparing to the electrical market by any RET.

In the **Figure 3** it is shown the changes in the net equivalent FIT, calculated from the on line data base of SNCE as a percentage of the total received retribution by the annual generated energy for each RET. If the functions are adjusted to a logarithmic tendency line, it is confirmed that the percentage of the equivalent FIT relative to the total received retribution from the different studied RETs tends to acquire in short terms values near 38.4% in the case of SH; 39.6% in wind energy; 50.9% in biomass, and 87.8% in solar PV.

From the **Figure 3** it can be figure out the following statement:

All the drawn lines as it is shown in **Figure 4** tends to increase with time, which involves low competitiveness in short term, although the biomass has the highest slope that means a bigger gap with the electrical market price comparing to other RETs.

MARGINAL COST CURVE OF THE GLOBAL RENEWABLE ENERGY

The marginal cost curve of the global Renewable Energy, biomass, Solar PV, SH and wind energy included, has been traced as a function of the percentage of RET (quotient between RET and the sold total energy at electrical market) that has contributed to the electricity generation (x axis in **Figure 4**). This curve shows the evolution of the additional cost of the RET in percentage respect the total retribution by the sold total energy at electrical market price (y axis in **Figure 4**). All the parameters necessary for the calculation has been taking from the on line data base of SNCE. This curve represents a test of effectiveness to achieve the Spanish target (rising of the global TER) versus their efficiency in costs (increasing overcost of RET) [8].

Table 3. Net equivalent FIT per MW installed development (Thousands EUR/MW)*.

Year	Technology			
	Biomass	Wind	SH	Solar PV
2004	91.231	39.611*	69.465	255.734*
2005	87.524*	44.447	47.591*	281.763
2006	113.757	59.472	64.805	265.397
2007	157.475	55.913	63.601	271.203
2008	198.268	59.296	61.723	281.964
2009	275.132#	70.849#	102.947#	696.069#

#: it represents the maximums values;
*: it represents the minimums ones.

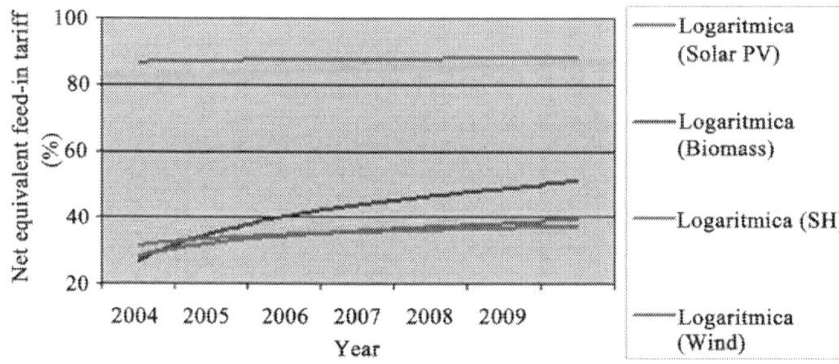

Figure 3. Net equivalent feed-in tariff for each RET.

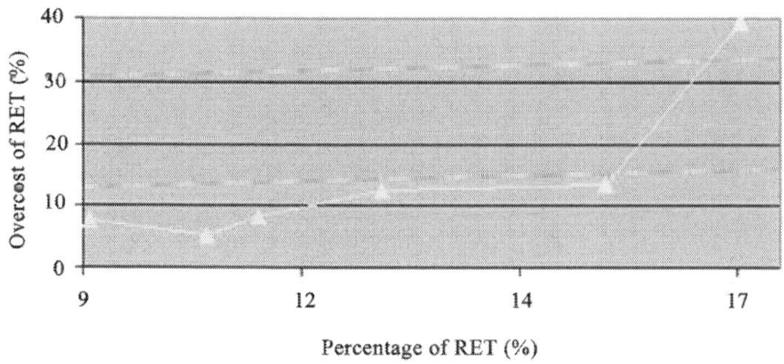

Figure 4. Marginal cost curve of RET.

The data represented in the **Figure 4**, belongs to the period from 2004 to 2009, both inclusive. From its profile it can be achieved two main remarks:

- The function grows respect the x-axis, which means a good approach to meet its annual target in Renewable Electricity Generation, that is to say, the policy to achieve the Spanish objective in RE generation is resulting effective.
- The last track of the curve, belonged to year 2009, entailed a huge increase of over cost remaining to the most expensive RET, Solar PV source, whose GWh sold in the electrical market it has been duplicated this year, therefore the policy to achieve the Spanish objective in RE generation is not efficiency in costs.

CONCLUSIONS

Conclusions can be categorized in three principal groups:

Principal Conclusions

The two main conclusions of the study are:

- The policy to achieve the Spanish objective in RE generation is resulting effective but not efficiency in costs.
- In short term it is not any RETs that can achieve competitive price comparing to the electrical market, particularly the biomass has a bigger gap with the electrical market price comparing to other RETs.

National Context

The variable working hours per MW is a curve almost symmetric for the SH, decreasing smooth for biomass and wind and decreasing sharply for the Solar PV. The decreasing slope can be caused by

- Lack of planning between the increasing development of the RETs power and the decreasing sold working hours per MW covered by the FIT system.
- Gradual reduction in the technological potential in the RET.
- Problems to adjust the energy generation and the electrical demand.
- Decrease in the electrical demand.
- Punctual fall in the renewable resources.

In the case of the Solar PV, its high FIT represents an unsustainable developing of this RET.

AACC Context

SH is the technology which development is raising the most among all the TERs and Solar PV is decreasing in the majority of AACC. We can also establish the following conclusions particular:

- In the period 2007 - 2009, the Solar PV has grown exceptionally in Canarias (234.0%) and Extremadura (326.1%), and the wind in Cantabria (4.5%). It is also singular that SH has decreased 68.0% in Canarias.

Félix Hernández, Miguel Hernández-Campos • In Navarra, the development of SH, Solar PV and wind has decreased harmoniously, becoming a rational initiative for the legal authorization granting a regulated utility to construct a RETs power plant.

NOTES

[1]In Spain technologies of special regime charge a FIT in the electricity price and include all RET and cogeneration, wastes and treatment of wastes. The latter is excluded of the RET, as opposed some EU country.

[2]Wastes and waste treatment are not considered RET by the Spanish law. However, both of these are included in the RETs in some of the EU member countries.

[3]Equivalent FIT is defined in the RD 485/2009 of the fourth of April of 2009 as the difference between the regulated tariff established and the settlement carried out for the sale of energy generated from renewable energy.

[4]European Commission [5] gives a relation between the capacity and the investment cost for biomass (1,124-1,406 EUR/MW); SH (1,700 EUR/MW) and wind energy (900 EUR/MW).

[5]In the legend of the Figure the y axis on the left side is the reference of the RET followed by PA in brackets, and on the right side when the RET is followed by SA in the brackets.

[6]SH feedback is due to the abundance or lack of this resource each year, which defines the framework of sold energy per MW installed. It has an unpredictable short term behavior.

[7]This is a common problem in Spain regarding to construction permissions provided by the Autonomic Community of new Power Plants, however, the regulation of the FIT system is leaded by the National Administration.

[8]A recent monitoring report from the SNCE shows many irregularities related to the working hours recorded in the Solar PV plants.

[9]As it has been said before, the information provided from the SNCE is reported until March 10th, 2010.

[10]No data is recorded from 2007 in this AC.

[11]The tons of CO_2 avoided have been obtained per each year from the energy mix. Source: www.idae.es.

[12]The total amount of the emissions avoided has been calculated by the average prize of the EU emissions market in the latest two years in EUR/Ton of CO_2 (Source: Electric and Business System of CO_2 Emissions RightsSENDECO2).

REFERENCES

1. P. Komor and M. Bazilian, "Renewable Energy Policy Goals, Programs and Technologies," Energy Policy, Vol. 33, No. 14, 2005, pp. 1873-1881. doi:10.1016/j.enpol.2004.03.003
2. U. Wang, "Spain: The Solar Frontier No More," Greentech Media, May 2009, accessed at 2 June 2009. www.greentechmedia.com
3. T. Couture and Y. Gagnon, "An Analysis of Feed-in Tariff Remuneration Models: Implication of Renewable Energy Investment," Energy Policy, Vol. 38, 2010, pp. 955-965. doi:10.1016/j.enpol.2009.10.047
4. CNE, "CNE Renewable Energy Statistics," 2010. www.cne.es
5. European Commission, "Sustainable Energy Technology Reference Information System (SETRIS)," Joint Research Centre European, 2004. www.jrc.es
6. J. Anta "El Mix de Generación Eléctrica a Future," 2010. www.energiaysociedad.es
7. L. Neij, "Use of Experience Curves to Analyse the Prospects for Diffusion and Adoption of Renewable Energy Technology," Energy Policy, Vol. 25, No. 13, 1997, pp. 1099-1107. doi:10.1016/S0301-4215(97)00135-3
8. P. Del Río, F. Hernández and M. Gual, "The Implications of Kyoto Project Mechanisms for the Deployment of Renewable Electric In Europe," Energy Policy, Vol. 33, No. 15, 2005, pp. 2010-2022. doi:10.1016/j.enpol.2004.03.022

17

The Development of Electricity Grid, Smart Grid and Renewable Energy in Taiwan

Hwa Meei Liou

Graduate Institute of Technology Management, National Taiwan University of Science and Technology, Taipei, Taiwan

ABSTRACT

The grid has played a vital role in the evolution of the electricity market; from traditional to smart grids; from fossil fuel power generated electricity grid connections to the integration of other renewable energy forms such as solar and wind power; the grid has played a key role in each step in Taiwan's move towards energy transition. This study includes Taiwan's construction of its transmission and distribution network, the recently passed newly revised version of the Electricity Act with its revisions to its transmission and distribution related content, and policies promoting the smart grid as well as issues that the renewable energy grid has raised in both the technical and legal aspects. Taiwan's electricity supply system is made up of the northern,

Citation: Liou, H. (2017), "The Development of Electricity Grid, Smart Grid and Renewable Energy in Taiwan". Smart Grid and Renewable Energy, 8, 163 177. doi: 10.4236/sgre.2017.86011.

central and southern systems. The Transmission and distribution grid have been defined as a common carrier, maintaining state-owned monopoly. The smart grid has 6 main facets to promote, including smart generation and dispatch, smart transmission, smart consumers, smart grid electricity grid industry and the establishment of a smart grid environment. Due to the possible effects of the integration of renewable energy generated electricity, there is a vital need for the regulation of the grid's management and skills.

Keywords: Electricity Grid, Smart Grid, Renewable Energy

INTRODUCTION

From traditional electricity grids to smart grids, from power grids based on fossil fuels such as coal and petroleum to grids integrating renewable energy sources such as solar and wind power, the grid system itself plays a key role in the operation of the electricity market, revealing the evolution process of this market, in particular in light of the problems which climate change and global warming raise for us now and for countries around the world; these issues are ones that will have to be addressed as we face the future. Energy saving methods can be achieved through advancements in technology and the development of new forms of power grids, while improving the production of renewable energy and reducing emission levels of greenhouse gases; those are all key links in enabling Taiwan to move towards energy transition.

A trend towards liberalization of the power industry took place around the world in the 1980s, and this reformation of the power industry included the transmission and distribution line within the electricity industry. With power grids aging and the demand for power increasingly rising, the importance of investing in the both the broadening of the scope of current power grids as well as the maintenance becomes more obvious day by day. The installation of smart meters is able to manage demands and is also an important part of power grids modernization process.

From a legal perspective, the need for legislative amendments to support this process has been shown to be a vital factor in power industry reforms in past research. This includes updating legislation with amendments to both the management and technological aspects supporting new developments, while revising outdated legislation to keep up with the progress of leading countries worldwide. The purpose of this paper is in discussing the development of Taiwan's distribution and transmission power grid, the current status of attempts to promote a smart grid, and the affects of the newest revisions to power industry legislation on the scope and content of transmission and distribution as part of the electricity market's liberalization policy plans, as well as the effect of renewable energy integrated grid on traditional power grids and the legislative response needed. Finally, this paper will consider the development of power grid related energy policy and legislation in the US and Europe and discuss possible ways that such developments could help Taiwan as it continues to promote renewable energy development.

TRANSMISSION AND DISTRIBUTION NETWORK

A country's power grid is part of its basic infrastructure, and the development of such power grids has a history of more than a hundred years [1] , power systems include electric power generation, transmission, distribution and sale systems, and within this system, transmission and distribution are formed through a combination of transmission and distribution grid paths and power substations. Taiwan's power grid system is run by state owned Tai-Power, and can be separated into three main systems, the north, central and southern systems, respectively based in Xinzhu County, Fengshan River and Choshui River. Figure 1 and Figure 2 show the northern, central and southern electricity systems' power supply capacity and peak load of power utilization in 2015. We can see from these figures that both power utilization in the north and the northern system's supply level are higher than the central and southern systems, with utilization accounting for 39% while supply levels dropped to 34% meaning that the north is already reliant on the economic dispatch of power from the central and

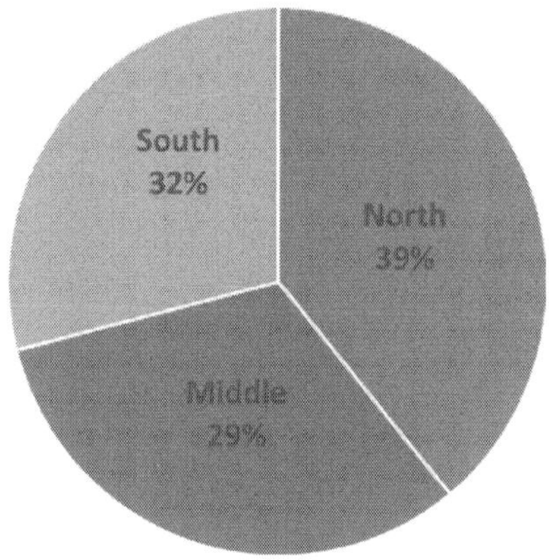

Figure 1. Peak load of northern, central and southern Taiwan.

Source: Taipower, available at: http://www.taipower.com.tw/content/new_info/new_info-c21.aspx?LinkID=12

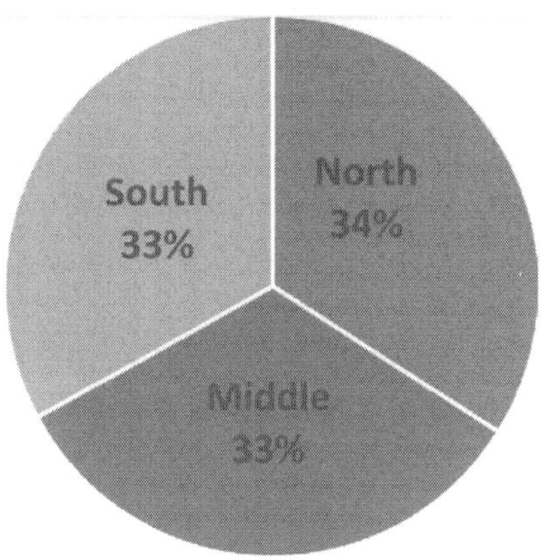

Figure 2. Load capability of northern, central and southern Taiwan.

Source: Taipower, available at: http://www.taipower.com.tw/content/new_info/new_info-c21.aspx?LinkID=12

southern systems to the north. For optimum results both the generation of power and power use should occur identically within the same system; the central system's generated power accounts for 33% of all power generated nationally, a figure higher than the 29% peak load the central system consumes. While the southern system's generated power also stands at 33%, slightly more than the system's peak load of 32%. Only the northern system produces less power than its peak load consumes, leading to a situation where the northern system lacks power. As for why the northern system lacks enough power, the main reason is the densely populated nature of the area, leading to higher demands for power than in central and southern Taiwan. Taipower's Power Development Plan continues to emphasize the regional balance of power in order to decrease the trend of power being transmitted from the south to the north [2] .

The process of power transmission is dependent on PSA levels, with the power system being dividable into three types, as Figure 3 shows: 1) Nuclear power, large scale hydropower and thermal power plants, the power produce from this type of plant having gone through a transformer produces 345 KV, making use of transmission lines to transmit energy and having passed through the power substation voltage is reduced to 161 KV, 69 KV, providing large scale users in science parks, before going through a power distribution substation to reduce voltage again for use in normal homes and civilian electricity; 2) Medium scale hydropower, thermal power and large scale renewable energy power plants; 3) Small scale hydropower and medium scale renewable power plants [3] .

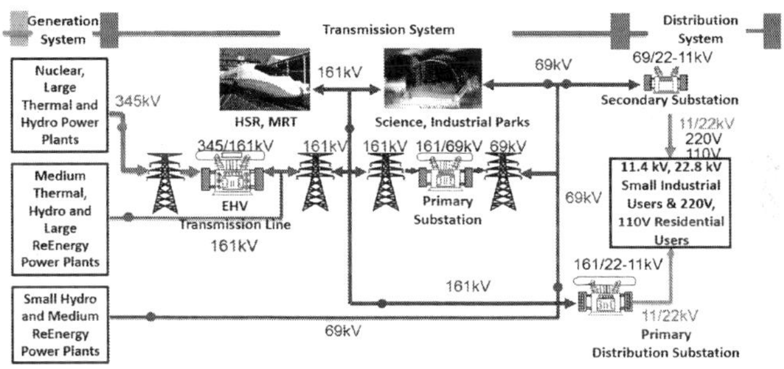

Figure 3. Introduction of Taiwan electricity supply system.

Source: Taipower, available at: http://www.taipower.com.tw/content/new_info/new_info-c21.aspx?LinkID=12

The power transmission and distribution line can be divided into three types: 1) 345 KV voltage-level extra high voltage (EHV) transmission power line, Taiwan has three such lines, using large quantity of electric charge, long distance transmission and 4004 KM loop; 2) 161 KV level primary transition line; 3) 69 KV level secondary transition line. The latter two use regional medium to small sized systems as a backbone with large power systems secondary networks, with a total of 13,281 KM loops [2] [4] . Distribution systems are the broadest link in the electricity grid, taking into account the size of the area and distinct nature of the distribution area, the density of load, system voltage and other factors bring out different patterns [5] .

Power grids can be divided into Super high Voltage (SHV), High voltage (HV), Low Voltage (LV) power grids [4] . Power transmission systems include 31 SHV substations (345/161KV), 45 primary transmission substations (161/69KV) and a distribution system including 234 distribution substations (161/22, 11 KV) and 293 secondary substations [3] . (Figure 3)

Taiwan's first power grid with the first transmission line was completed in 1951, connecting Hualian to the West coast's electricity system; in 1962 the Hualian Taitung secondary transmission line was completed, connecting Taitung to the west coast electricity system; in 1974 the first 345 KV SHV transmission system line between North and South was completed, at the time Taiwan was the first country in Asia to have a 345 KV transmission system [6] .

Taipower's organizational structure includes a power generation division, transmission system division and distribution and service division, in charge of generating, transmitting and distribution of power [7] .

There are two ways to construct power transmission and distribution lines: the first is overhead lines, constructing transmission towers, cement poles, with one tower connected to the next tower covering the whole of Taiwan. The problem with this method is that it makes use of more land, making it easy to give rise to dispute. Another method is ground cables, building pipelines underground, along with cable conducting installation. In 1974 Taiwan began building Taiwan's first HV underground cables [6] .

REGULATIONS ON TRANSMISSION AND DISTRIBUTION NETWORKS FROM THE PERSPECTIVE OF THE NEWEST REVISION OF ELECTRICITY ACT

The liberalization of Taiwan's Electricity Industry planned to begin in 1995 when the Executive Yuan first sent the draft revision of the Electricity Act to the Legislative Yuan for deliberation, then after in 1999, 2002, 2007, 2008, 2015 and 2016 new amendments to the Electricity Act were sent to the Legislative Yuan. Currently, the newest version of the Electricity Act was revised in January of 2017, in particular with Taiwan having a new ruling party, the Bureau of Energy, Ministry of Economic Affairs once again proposed a new version in 2016. Revisions to the Act were accessible on line, and then gradually convened to discuss, consulting related administrative agencies, civil groups, electricity industry and labor groups to understand various opinions.

The latest Revision of the Electricity Act has 9 chapters, and 97 articles and the revisions made regarding regulations on Transmission and distribution networks are as follows: (please refer to Revision of the Electricity Act)

1) *In the future the power industry will be divided into three distinct industries:* power generation industry, transmission and distribution industry and electricity selling industry (article 2 item 1), each with its individual management.

Taiwan's current electricity market structure is based on one state-run integrated power company combining the business of the generation, transmission distribution and selling of electricity. While the electricity generation department is open to independent power producers (IPP), combined heat and power station, renewable energy power generator, yet besides Taipower, the electricity produced by other sources, legally speaking are only permitted to sell wholesale to Taipower, and are not allowed to sell directly to the consumer. That is to say that Taipower currently holds the exclusive right to the operation of the electricity business (old electricity act article 3). In the future users will not be limited to purchasing electricity from Taipower alone. Electricity industry organizations, other than renewable energy generator plants, will be limited to Company Limited by Shares.

2) The transmission and distribution sector will still remain a monopoly

"The transmission and distribution sector is defined as" points to the national installation power grid, providing power for public use (article 2 item 4), "the electricity grid is defined as refers to the demarcation point between the main electricity power generating equipment, the transmission and distribution industry and the user, all belonging to the same electricity transmission system with its supporting installations and substation installation." (article 2 item 14) Since the transmission and distribution network possesses both public and natural exclusive properties, by defining the transmission and distribution power industry equipment as "public access" or "common carrier", does the Act maintain the only one state-owned enterprise. The transmission and distribution industry's wheeling of renewable energy power generator's electricity, or renewable energy power generator's can customize an in-house route to directly transmit electricity to users.

3) Transmission and distribution power industry to operate power transmission and distribution network

The transmission and distribution power industry will be responsible for the national power transmission and distribution network, including the planning, construction and maintenance, with an obligation to ensure users are connected, while fulfilling their obligation to ensure interconnecting grid systems for other power industries. It is vital to ensure the fairness of the power grid, equal treatment for all and public provision for all power industry users. This version of the legislation has yet to require the establishing of an independent systems operator (ISO) for the transmission and distribution sector.

4) New organizational design

To coordinate with the development of the transmission and distribution industry, the new organization will have follow qualities. First, having referred to the experience of the UK, US and EU, an Electricity Industry regulatory agency would be assigned by the central competent authority in charge of the state- owned power related business. Secondly, a Platform for Power Trading would be set up by the Transmission and Distribution industry.

5) Concurrent limitations

There are two main limitations: 1) The operation of a transmission and distribution company, in principle, can't operate concurrently in terms of the power generating industry and the electricity selling industry; 2) The transmission and distribution industry supplementing other industries besides the power industry, must not influence its own business operations, nor should it limit competition or cause unfair competition, moreover it should be regulated by the Electricity Industry Regulatory Agency.

As for the power generation industry and normal Power Sales industry, since they count as non-public businesses, they can work in a free competitive market and diversify their businesses, unlike the old Electricity Act with its concurrent limitations.

6) Terms of license permits

The terms of license permits for the electricity industry are all the same, 20 years, once the license permit term comes to an end, the business can apply to the Electricity Industry Regulatory Agency for an extension.

7) The establishment of a power development foundation

This revision of the Electricity Act aims to set up a dry fund, in relation to the transmission and distribution industry: the transmission and distribution industry and power generating industry should invest a lump sum to establish a power development foundation, with the legal purpose of taking into account businesses' social responsibilities and feeding back to the community and residents of areas around power plant installations.

8) 2025 achieve a nuclear-free homeland

Taiwan currently has four nuclear power plants, the building of the 4th has already been brought to a halt, and there will not be an extension to the life of power plants' 1st to 3rd. This revision sets 2025 as a goal for the end of operations at all nuclear plants. Within this context, in order to make up for the lack in the power supply which the end of nuclear power in Taiwan will lead to, it is of vital importance to be proactive in developing forms of renewable energy and strengthening the construction of smart grids to enable the country to cope with the changes that will occur in the energy resources.

In Taiwan's promotion of the liberalization of the power industry, there should be an increase in competition for the power generation market; increasing business operations efficiency and bringing about growth in power production industry. At the same time the renewable energy industry would be developed, increasing Taiwan's image as a country that promotes green energy and low carbon society within an international community. Fitting this in with the long term plan for promoting the smart grid we can see that it would be contributions to the economy and society as a whole.

SMART GRID

The concept of developing Smart Grids was one of the main aims of the Executive Yuan's 2010 National Master Plan on Energy Conservation and Carbon Reduction. Beginning in 2009 with the National Energy Conference held by the Ministry of Economic Affairs, the developing of smart power meters and smart grids was set as a principle item for forthcoming promotion, including electrical terminal of the electricity transmission grid, power distribution all the way along to the user clients smart meter, smart grid and this has been written in as one of the principle axis of the National Energy Program for which the Ministry of Science and Technology is responsible, smart grids are also an important bridge to be established to enable greater interaction between the energy industry and information and communication industry. In 2010 the Executive Yuan passed the "Advanced Metering Infrastructure", AMI and in 2011 the Ministry of Economic Affairs Energy Board established a Working Group for the Smart Grid Principle Central Plan, as well as holding a Smart Grid Development Strategies Forum. In 2012 the Executive Yuan passed a Master Plan of Smart Grid in Taiwan, becoming an important policy in the promotion of smart grids, moreover establishing a cross-sector Interdepartmental Promotion Team of Smart Grid, while Tai Power also established a smart grid working group.

According to the definition used in the Master Plan of Smart Grid in Taiwan: "Through data, information and automatic technology, construct an integrated electricity grid integrating power generating, transmission and distribution, enabling automatic, secure close coordination between clients and the supply side, promoting greater efficiency in the electricity system's integrated transportation, providing quality and reliable power grid, while promoting the expansion the use of renewable energy and carbon reduction policy goals". And this smart grid can be divided structurally into 6 different aspects, including: Smart Generation and dispatch, smart transmission, smart distribution, smart consumers, smart grid industry and constriction of a smart grid friendly environment. It was estimated in 2011 that the smart grid's development would be separated into three stages, over a 20 year promotion period, these three stages are as follows: Progressively Extending Stage (2011-2015); Promoting Stage (2016-2020), Expansion Use Stage (2021-2030) [8] [9] , as shown in Figure 4. An estimated 139.9 billion NTD will be invested into this project; this will be invested correlating with each different stage as shown in Figure 5.

The definition of a smart meter, according to Brendan Cook et al.: A smart meter is a device which monitors a household's electricity consumption in real-time, and has the ability to display real-time pricing in each household [10] . A smart meter is capable of distinguishing between high voltage AMI and low voltage AMI. Between 2010-2012 the Ministry of Economic Affairs Energy Bureau completed the installation of a smart meter model system (Taipei, Xinzhu, Tainan), by the end of 2015, all 24,000 high voltage users, which accounts for 60% of power utilization in Taiwan, and 10,000 low voltage users, had already installed smart meters [11] .

Taiwan's first Smart Grid Demo Site and AMI is currently installed in Penghu [12] , There are other demo sites such as smart user energy management system demo sites in a number of places including Tainan, Academia Sinica, convenience stores, science parks, China oil company, along with smart power distribution system technology and virtual power plant technology in various areas and institutions including Tai Power, the Business sector and Jinmen, altogether there are 18 smart grid technology demo sites, installed by the Ministry of Economic Affairs, as well as the technology and business sectors [4] [13] .

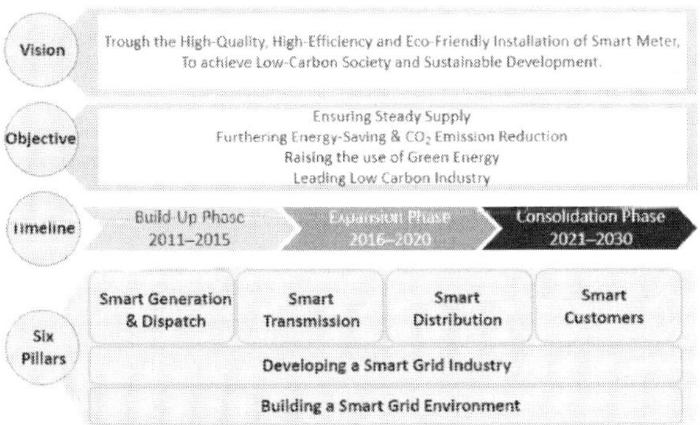

Figure 4. General Framework of Smart Meter.

Source: Bureau of Energy, Ministry of Economic Affairs, R.O.C., General Program of Smart Grid

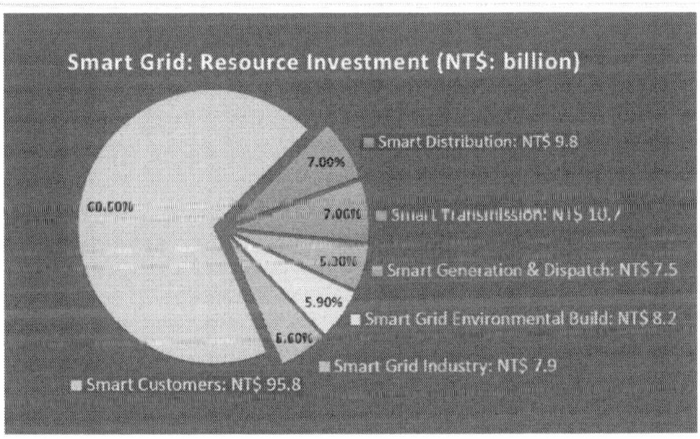

Figure 5. Smart Grid: Resource Investment

Source: Bureau of Energy, Ministry of Economic Affairs, R.O.C., General Program of Smart Grid.

In terms of important plans for developing smart grid use, the National Energy Program—Phase II, included a "smart grid major central plan"[1] [14] . In terms of the smart grid industry there are currently two important associations: the Taiwan Smart Grid Industry Association and Taiwan Smart Energy Industry Association. In terms of the market, the market for building grid foundations can be divided into three main kinds: electricity wiring, substations and control facilities and terminal facilities [4] .

In terms of regulations regulating the construction of smart grids, including the Electricity Act, Grid Interconnection Code of Renewable Energy, Regulation of Exterior/Interior Circuit Installed [15] , and others such as the Energy Management Act, Renewable Energy Development Act are all related in some way [16] , besides this there are also related issues such as information security, privacy rights that have related regulations [17] [18] , intellectual property rights related regulations, building standardization and authentication systems and so on related standards [19] .

Under the larger energy policy goals for the promotion of renewable energy, smart grids play an irreplaceable role, as smart grids enable the promotion of large scale generation of renewable energy into the grid. Smart grids combine smart meter, enabling management of demands, reducing CO_2 emissions [4] . Through implementing smart grids, electricity systems will be able to ensure high level operations, improving the quality of electricity provision and systems security [20] . As for the development and improvement of smart grids nationally, this is affected by the promotion policies of the government that are dependent on a number of factors including the progression of power electronics, smart grid information and smart meter [21] .

As for South Korea's legislation related to promoting smart grids, in 2011 the Smart Grid Promotion Act was passed providing a sustainable smart grid development planning framework. While in the aftermath of Japan's Fukushima nuclear disaster the adoption of smart meters helped to manage demand. Under this context the Ministry of Economy, Trade and Industry (METI) continued to promote the construction of smart grids [22] . While in China smart grids are listed as part of the 12th Five-Year Plan as an important industry in the strategic development of new industries, the planning for which is separated into seven main areas: generated power, transmission of power, substations, allocation of power, power utilization, scheduling and communication platform [8] . In Taiwan smart grids were included in the energy policy as a priority for promoting in 2009 and now have already entered the second stage, though as yet no special law has been formulated.

INTEGRATION OF RENEWABLE ENERGY INTO THE GRID

Due to the market penetration rate and intermittent characteristic of renewable energy, once large amounts of renewable energy is incorporated into the grid, it will have a direct affect on the quality of electricity provision and reliability [23] . Many countries when allowing renewable energy to be connected to the grid have provided standards for the management and technology of such challenges including variation in voltage, frequency etc.; in order to ensure the flexibility of grid transmission once renewable energy is connected to the grid, standards were also put into place : voltage regulations, information data etc.; as well as grid operation, real-time pricing are all aspects which can be regulated through establishing standards to respond to the challenges that the connection of renewable energy to the grid could give rise to [24] . In terms of the extent to which legislation covers these issues, amendments have been made to current legislation to respond to the newest development trends in the electricity grid, as well as making new laws to regulate the future of electricity grid's modernization, and the challenges that smart technology could give rise to. Take for example Germany's Power Grid Expansion Act, EnLAG.

The key legislation for Taiwan's implementation of renewable energy being incorporated into the grid is the Renewable Energy Development Act Item 8: "Renewable energy power facilities and the generation of electricity power shall have the stability of their power grids evaluated by local electric power grid enterprises and have them paralleled and bought wholesale at the locations where existing power grids are closest to renewable energy power assembly sites and provide electricity required by such power facilities during maintenance shutdown period; electricity enterprises shall not reject the aforesaid request without proper reason and approval of the central competent authority." In particular this emphasizes the sharing of costs: other than the existing circuits, the cost for power grid enhancement will be shared between both the electricity enterprises and renewable energy power facility installers. The rising cost for the enhancement of the power grid is for example the expansion of current lines, increasing related transformers, or user line being changed to larger scale [25] . Moreover, in terms of line costs: The installer shall install and maintain the circuits connecting renewable energy power facilities and power grids. The renewable energy power generation facility installers shall bear the costs incurred.

Besides the regulations mentioned above, other legislation related to grid- connection also include standards related to electricity: for example: regulations related to electric utility circuit installations) (originally known as interior electricity supply power lines installation regulations, Ministry of Economic Affairs Standards); regulation of interior circuit installations) (Ministry of Economic Affairs Standards), along with business regulations as set out by Taipower: for example Taipower's Guide to Renewable Energy Electricity Generated System Grid Connection Integration Technique, Taipower's Guideline for energy distribution system planning, Third type (installation capacity less than 500 KW) Renewable energy self-use Electricity generating equipment parallel connection framework [25] . Renewable energy electricity generating system grid-connection technique framework regulating renewable energy grid-connection power distribution system electricity voltage level and capacity limits, stipulating renewable energy equipment grid-connection protection support plan, design and installation standards along with grid-connection transmission demands [26] ; Besides this, Taipower established the renewable energy electricity generation grid- connection capacity inquiry system, enabling users to check grid-connection capacity for their area with the intention of applying for grid-connection.

In terms of the constructing of the electricity grid, research points out that legislation's inability to coordinate with developments is the largest barrier against policies attempting to promote the extension of the electricity grid, with an influence far greater than that of the development maturity of electricity grid technology. This Legislation barrier includes: permission procedure delays, the procedure itself being too complicated and lacking transparency; nationally, there are a lack of well planned, long-term guidelines for the expansion of grid; internationally, there is also a lack of planned cooperation. The issues mentioned above show that there is a failure on the part of both the standard system content and implementation process. Sometimes the current standards themselves might be the greatest obstruction to development. In order to eliminate this legislation barrier to development, rather than pointing out small adjustments to be made within the narrow range of current legislation in terms of revisions to specific content or improvements to the legal body at an implementation level, there is a need for significant changes to be made to the regulatory framework itself [27] .

US AND EU DEVELOPMENT

The US legal system and the legal process in relation to smart grids, is closely related to the development some bills related to US electricity market [28]: In 1965, the northeast experienced its first power failure, in 1977 the Federal Power Commission (FRC) reformed as the Federal Energy Regulatory Commission (FERC), then in 1978 the National Energy Act was passed, which then led to the Public Utilities Regulatory Policies Act (PURPA). In 1992 the Energy Policy Act was passed and then in 2003 Northeast once again experienced a power failure, following this in the years 2005, 2007 and 2009 the 2005 Energy Policy Act, Energy Independence and Security Act (EISA07) and the American Recovery and Reinvestment Act (ARRA09) were each passed respectively. Of these Acts the Energy Independence and Security Act directly dealt with Smart Grids and added related regulations, in Title XIII specified the "Statement of policy on modernization of electricity grid", a "Smart grid advisory committee" and "Smart grid task force", "Smart grid technology research, development, and demonstration" and so on. Then the American Recovery and Reinvestment Act set a budget to invest in the development of Smart Grid, including Smart Grid Investment Grant Program, Smart Grid Demonstration Program [29] . In terms of policy, in 2003 the U.S. Department of Energy proposed the "'Grid 2030': A National Vision for Electricity's Second 100 Years", in 2012 it proposed the "2010 Smart Grid Report" [30] .

As for EU, their smart grid legislation is related to climate change and energy policy: the EU's 20-20-20 strategic goals and the Strategic Energy Technology Plan (Set-Plan) and other related initiatives such as the European Electricity Grid Initiative (EEGI) the plan of which is part of the Set-Plan. In 2005 the EU established the Smart Grids European Technology Platform; in 2009 the Third Energy Package goal was to complete the Internal Energy Market. The EU's R&D Framework Programme (FP6, FP7) all support smart grid development, according to the newest R&D Framework Programme (Horizon 2020), between 2014-2020 the EU will invest an estimated 80 billion euros into research innovation plans, on a theme called "Societal Challenge" directly related to smart grid and referred to as Secure, clean and efficient energy [31] . The EU's Electricity Directive demands that member states develop a smart grid system plan in compliance with the directive and establish a time frame for its development by the 3rd September 2012 [32] . Other related legislation include the Energy Infrastructure Package, the Regulation on Energy Market Integrity and Transparency and the Energy Efficiency Directive. In 2011 the EU Committee proposed the Connecting Europe Facility (CEF) legislation package, with goals for improving energy, transport, information these three connected infrastructures [33] . The future development of the EU's electricity grid will be closely related to whether or not they achieve the 20-20-20 goals.

CONCLUSIONS

As a result of the rapid development of renewable energy, advancements in digital technology and the growing connection between energy and information technology, the electricity grid is currently in a period of reform. And while this could benefit aims to ensure energy saving and improvement of energy efficiency, reliability and security are still the electricity market's major principles within the electricity provision process.

Taiwan's electricity supply system is actually made up of three systems: the north, central and south systems, of these systems the northern system's supply and demand are the highest. The Transmission system includes high voltage substations and primary substations; power distribution systems include power distribution substation and secondary substations; altogether Taiwan has three high voltage transmission electrical lines. Currently the electricity market is dominated by the state-owned Taipower; however the liberalization of the power industry is currently under way and is already in full spring. In January of 2017 the Revision of Electricity Act is passed. Transmission and distribution grid have been defined as a common carrier, maintaining state-owned monopoly, not opening up to competition. The Amendment plans for Taiwan to become a nuclear free country by 2025.

As for the Smart meter, in 2010 the Executive Yuan passed the Advanced Metering Infrastructure, (AMI) plan; currently high voltage users' installations have already fulfilled the quota set in the AMI plan, while low voltage users installations are on track for reaching targets. As for the Smart grid, policy is mainly based on the "Smart Grid Principle Central Plan" passed by the Executive Yuan in 2012, and is separated into six facets, including Smart Generation and dispatch, Smart Transmission, Smart distribution, Smart Consumers, Smart Grid Industry and Smart Grid Environment Construction. Beginning in 2011 the promotion of Smart Grid has been separated into three stages; and currently development has reached the stage of having built demo sites, accumulating experience and persisting in pushing forward, in terms of national plans; the development of smart grid is being pushed through the Smart Grid Major Central Plan. Renewable energy grid-connection is mainly based on Renewable Energy Development Act article 8, related to strengthening electricity grid pricing, as the current legislation states that renewable energy generators and electricity generating enterprises should share the burden. Other legislation influencing the grid-connection issue, is the Ministry for Economic Affairs, the electricity related executive directive and Taipower company's business regulations, Taipower has also set up a renewable energy electricity generator grid-connection capacity inquiry system, which enables one to directly confirm the actual grid- connection capacity. In order to respond to the intermittent characteristic of renewable energy, when connecting to the grid, it will have a direct affect on the stability and reliability of the grid; therefore there is a need to learn from the US, EU and other countries experience and methods in developing grid-connection techniques, management methods and general standards.

REFERENCES

1. Chang, C.A., Lan, B.R., Lin, Y.J., et al. (2014) Challenges in the Development of Smart Grid and Renewable Energy. Journal of Taiwan Energy, 1, 259-281.
2. Zeng, J.-R. (2015)The Transmission of Electricity from the South to the North. Energy Monthly.
3. Taipower Website (2016) http://www.taipower.com.tw/content/new_info/new_info-c21.aspx?LinkID=12
4. Taiwan Smart Grid Industry Association Website (2016) Understanding the Smart Grid. http://www.smart-grid.org.tw/
5. Wang, Y.C. and Huang, Q.T. (2014) The Development and Challenge of Smart Distribution Grid. Journal of Electrical, 168, 76.
6. Zhu, R.Y. (2010) The Transmission Line System of Taiwan Power Company Which Covers the Whole of Taiwan. Yuan Magazine, 79, 5-6.
7. Taiwan Power Company (2015) Taiwan Power Company Sustainability Report. Report, Taiwan Power Company, Taiwan.
8. Bureau of Energy, Ministry of Economic Affairs (2012) The Smart Grid Master Plan. http://web3.moeaboe.gov.tw/ecw/populace/content/SubMenu.aspx?menu_id=1946
9. Bureau of Energy, Ministry of Economic Affairs (2012) Program of Smart Grid Promotion Workforce. http://web3.moeaboe.gov.tw/ecw/populace/content/SubMenu.aspx?menu_id=1946
10. Cook, B., et al. (2012) The Smart Meter and a Smart Consumer: Quantifying the Benefits of Smart Meter Implementation in the United States. Chemistry Central Journal, 6, S5. https://doi.org/10.1186/1752-153X-6-S1-S5
11. Bureau of Energy, Ministry of Economic Affairs (2015) Energy Statistics Year Book. http://web3.moeaboe.gov.tw/ecw/populace/content/SubMenu.aspx?menu_id=1946
12. Chen, X.X. (2015) Plan for Smart Grid Technology Applications. Energy Monthly, 11, 34.
13. Chen, Y.H. (2014) The Current Status and Prospects of Taiwan's Smart Grid Industry. Taiwan Economic Research Monthly, 37, 17-18.
14. National Energy Program Phase II office (2014) The Introduction of National Energy Program-Phase II. http://www.nepii.tw/language/en/

about-nep-ii/introduction/

15. Bureau of Energy, Ministry of Economic Affairs (2012) Briefing on the Smart Grid Master Plan. http://web3.moeaboe.gov.tw/ecw/populace/content/SubMenu.aspx?menu_id=1946
16. Lee, K.Y. (2009) A Discussion and Response to Research Analysis on the Issue of Key Legislation Related to the Smart Grid. Science & Technology Law Review, 12.
17. Su, W.X. and Lee, K.Y. (2012) Industry R&D Invests in Intelligent Energy for Privacy Protection and Information Security in Response to Suggestions—A Case Study of European Union Promoting Legal Policy. Science & Technology Law Review, 49.
18. Zeng, Y.X. (2016) A Brief Discussion of the Importance of My Country's Key Energy Related Infrastructure Information and Communication Technology (ICT) Security Legal Structure—An Example from the Progressive Development of the EU and Germany's Smart Meter. Science & Technology Law Review, 18.
19. Chen, Z.X. (2015) Countermeasure Research on the Trend of Smart Grid Technology Development and Renewable Energy Integration into the Grid. Journal of Taiwan Energy, 2, 148.
20. Brown, M.A. (2014) Enhancing Efficiency and Renewable with smart Grid Technologies and Policies. Futures, 58, 30. https://doi.org/10.1016/j.futures.2014.01.001
21. Hossain, M.S., et al. (2016) Role of Smart Grid in Renewable Energy: An Overview. Renewable and Sustainable Energy Reviews, 60, 1182. https://doi.org/10.1016/j.rser.2015.09.098
22. Tuballa, M.L. and Abundo, M.L. (2016) A Review of the Development of Smart Grid Technologies. Renewable and Sustainable Energy Reviews, 59, 710-725. https://doi.org/10.1016/j.rser.2016.01.011
23. Taiwan Power Company (2013) The Impact and Influence of Japan's Fukushima Nuclear Disaster on the Energy Policy and Direction of Reform in Taiwan's Power Industry as Well as the World's Major Countries. Report, Taiwan Power Company, Taiwan.
24. Chen, Z.X. (2015) Countermeasure Research on the Trend of Smart Grid Technology Development and High Renewable Energy Grid Operation. Journal of Taiwan Energy, 2, 146-147.
25. Chen, Y.C. (2015) Renewable Energy Integration into the Grid Operation Practices. Taiwan Power Company Tainan District Office,

Taiwan.

26. Wang, Y.C. and Huang, Q.T. (2014) The Development and Challenge of Smart Distribution Grid. Journal of Electrical, 168, 77.
27. Battaglini, A., et al. (2012) Perception of Barriers for Expansion of Electricity Grids in the European Union. Energy Policy, 47, 254-259. https://doi.org/10.1016/j.enpol.2012.04.065
28. Simões, M.G., et al. (2012) A Comparison of Smart Grid Technologies and Progresses in Europe and the U.S. IEEE Transactions on Industry Applications, 48, 1154-1162. https://doi.org/10.1109/TIA.2012.2199730
29. Huang, Y.S. (2013) Smart Grid Promotion Policy in the United States (Part 1). Electricity Monthly, 23, 167.
30. Huang, Y.S. (2013) Smart Grid Promotion Policy in the United States (Part 1). Electricity Monthly, 23, 164-165.
31. Crispim, J., et al. (2014) Smart Grids in the EU with Smart Regulation: Experiences from the UK, Italy and Portugal. Utilities Policy, 31, 88. https://doi.org/10.1016/j.jup.2014.09.006
32. Chen, Y.H., et al. (2012) An Introduction to the Promotion of Smart Grid in Key Countries. Electricity Monthly, 22, 8.
33. Crispim, J., et al. (2014) Smart Grids in the EU with Smart Regulation: Experiences from the UK, Italy and Portugal. Utilities Policy, 31, 87-89. https://doi.org/10.1016/j.jup.2014.09.006.

18

Evaluation of Renewable Energy Vulnerability to Climate Change in Brazil: A Case Study of Biofuels and Solar Energy

Antonio Oscar Jr., Wanderson Luiz Silva, Vera Ruffato, Renata Barreto, Marcos Freitas

Federal University of Rio de Janeiro, Rio de Janeiro, Brazil

ABSTRACT

This study aims to calculate indicators and indexes to subsidize the analysis of vulnerability and adaptation of the renewable energy sector to climate change in Brazil, focusing on biofuels and solar energy. For biofuels, in general, the Brazilian coast will be a propitious area for agricultural productivity during the XXI century, but these are areas historically intended for occupation and development of the urbanization process, that is, with limited land availability and supply for primary production. In some parts of the Northeast, Midwest and South of the country, offer for the cultivation land will be reduced. For the solar energy is observed that Brazil has area and highly expressive power for the use of this power, both today and in

Citation: Jr., A., Silva, W. , Ruffato, V., Barreto, R. and Freitas, M. (2015), "Evaluation of Renewable Energy Vulnerability to Climate Change in Brazil: A Case Study of Biofuels and Solar Energy". Smart Grid and Renewable Energy, 6, 221-232. doi: 10.4236/sgre.2015.68019.

the coming decades, especially in the North, Northeast and Midwest. In statistical terms, the Mann-Kendal test and Sen's Bend point to a very weak tendency to useful radiation indicator in all regions of Brazil by 2100. In addition, it is projected a significant increase in mean air temperature by the end of XXI century across the country that can mean a reduction in power conversion capability, which is sensitive to ambient temperature variations, especially in the Midwest and North of the country.

Keywords: Renewable Energy, Climate Change, Brazil, Biofuels, Solar Energy

INTRODUCTION

Climate change increasingly occupies an important position within the various sectors of society, both nationally and internationally. Several studies, including the fifth assessment report of the Intergovernmental Panel on Climate Change [1], indicate that the growing trend of greenhouse gases is the main agent of such changes. Thus, there is an imminent concern about the mitigation of these future climate changes, making it necessary to establish measures to adapt to possible impacts.

As the power generation and consumption are key to socio-economic development, it is essential to understand how the energy production infrastructure responds to the clash of climate change. From this knowledge is possible to measure the vulnerability of a nation and promote development of alternative means of it, offering appropriate prospects for the future. In Brazil, due to its characteristic of primary producer associated with continental proportions of its territory and climatic aspects, its energy matrix were guided primarily in renewable energy.

Energy sources exploited in Brazil include hydropower, thermal power, solar power, biofuels and wind energy. According to the National Electric Energy Agency [2] , the country has 3336 projects of electricity generation in operation, representing an output of approximately 118 Gigawatts. Thus, significant climate changes may represent substantial impacts for development and for the country economy in this sector. Specifically this research, the main focus will be on solar energy and biofuels, as these sources have significant enhancement conditions and vulnerability to present and future climate scenarios.

The energy originated from biofuels is from renewable sources which in Brazil are essentially represented by biodiesel and ethanol. For proper use of land with purposes of exploitation of agricultural potential for the generation of biofuels, in addition to economic and social factors, it must take into account basic environmental factors, considering the physicochemical requirements of each culture, as well as the ecology of the region. Among the physical and ecological conditions, the climate plays an important role, especially with regard to water availability, which depends directly on rainfall supply and intensity of evapotranspiration process, elements that act as limiter to vegetation.

For its intertropical location, Brazil has a high incidence of solar radiation throughout the year, making this a potential source for the exploration of energy in order to supply the increasing demand of energy in the country. Most of Brazil is located relatively close to the equator, so that there are not wide variations in daily sunlight. However, most of the population and socio-economic activities of the country are concentrated in more distant regions of equator, as in the Southeast and South.

In view of the above perspectives, this work presents a vulnerability analysis and interpretation of the impacts (both positive and negative) of future climate changes in the renewable energy sector of biofuels and solar energy, in order to support researches that aim to adaptations and implementations in the current Brazilian energy matrix.

METODOLOGY

Data from Climate Model

The Numerical Models of the Earth System (NMESs) are the most promising tools to develop projections of future climate change. Through the simulation of important physical and dynamical processes, NMESs can represent the complex nonlinear interactions that influence the climate as well as the interactions between the components of the Earth system (mainly atmosphere, biosphere and hydrosphere) and feedback mechanisms, including changes in the frequency and intensity of extreme events [3] . In addition, NMESs can simulate future climates in response to changes in the concentration of greenhouse gases and aerosols.

For this study, the regional climate was simulated by Eta model of Weather Forecasting and Climate Studies Center/National Institute for

Space Research (CPTEC/INPE), which is coming from the original Eta model [4] developed at the University of Belgrade (Serbia) and operationally implemented by the National Center for Environmental Predictions (NCEP) in the United States. The Eta 20 km regional model scenarios were simulated using the global model boundary conditions MIROC (Model for Interdisciplinary Research on Climate), a model developed in cooperation with the climate research center in Japan. From these integrations, the model is called Eta-Miroc5 [5] [6] .

In this study, the average of thirty years is used as reference model for the processing of data and evaluation of future scenarios of the indicators described hereafter. The RCP 8.5 scenario [1] was selected to perform the climate change analyzes. Historical values (1961-1990) and projected future values (2011-2040, 2041-2070 and 2071-2100) were calculated. Then the future-present differences were computed in order to verify the simulations of future trends.

Biofuels Indicator

Assuming that water availability is fundamental to the farm of genres used for the production of biofuels, including ethanol, it is used the aridity index (D) proposed by [7] :

$$D = \frac{1}{(0.01 * IM) + 1}$$

where $IM = 100 * \left(\frac{PRECP}{EVTP} - 1 \right)$ is the Thorntwaite Moisture Index, PRECP is the precipitation and EVTP is

the evapotranspiration.

From this method, it was shown that occurrence of areas of desertification more emphatic process, so those with very low agricultural potential, are in the range $2 < D < 7$.

To complement this methodology is also used as reference humidity data in the soil in 0.1 m, which aims to support the assessment of water availability to the root system of plants. Furthermore, air temperature is also analyzed, from which proposes an index based on its variation (projected temperature-historical temperature), allowing estimation of the thermal stress of the plant in climate change scenarios, fundamental factor that directly affects the photosynthetic capacity of the plant.

Solar Energy Indicator

Assuming that the use of solar energy depends on the incident solar radiation in a given area, an efficient method of analyzing the view of the offer of this source is the radiation balance in a given region, which can be determined by calculating the balance of radiation. This radiation (Rn) is the available energy in the earth-atmos- phere system and is defined as the radiation balance of all radiative fluxes leading to and from a surface [8] . Therefore, to define a useful radiation indicator (Ru) is required equating the variables of shortwave radiation incident to the surface (OCIS), albedo (α), short wave radiation emerging at the top of the atmosphere (ROCE) and short wave radiation emerging to the surface (OCES). Thus:

$$\mathrm{Ru} = \mathrm{OCIS} + \mathrm{ROCE} - \alpha \cdot \mathrm{OCIS} - \mathrm{OCES} \left[\mathrm{W/m^2}\right]$$

The albedo (fraction of incident radiation which is reflected) regarded was the planet mean value, so 0.30 or 30%. The maps show the average annual values of the indicator for a particular day of the year in Brazil, so Ruis multiplied by 24 hours, having a send unit $W \cdot h/m^2 \cdot day$.

The equipment controllers of solar energy can be markedly influenced by variations in ambient temperature. Thus, also it is performed an analysis of projections of future increase or decrease in air temperature along the XXI century in Brazil.

Statistical Tests

For the evaluation of significance in climatic trends, it is used the Mann-Kendall nonparametric statistical test [9] . This test has been the most appropriate method to analyze the significance of possible climate change in time series [10] .

Sen's Bendis also calculated in this work. This is a non-parametric method (assuming a linear trend) used to estimate the trends magnitude [11] . As Sen's Bend is insensitive to outliers and missing data, such testis more rigorous than the usual curvature of regression and provides a more realistic measure of trends in a time series.

RESULTS

Air Temperature

Figure 1 shows the differences in air temperature between the future climate (2011-2040, 2041-2070 and 2071- 2100) and the historical period (1961-1990) projected to Brazil through Eta-Miroc5 model in IPCC RCP 8.5 scenario. The estimates show an increase of about +2.0°C in Brazil during 2011-2040. During 2041-2070, the projections indicate a similar temperature rise, but more sharply in the center of the country, with values reaching up to +3.0°C in the Midwest region and vicinity. Analyzing the period 2071-2100, it is noted that the increase in air temperature is emphatic in Brazil. The simulations show an increase between +2.0°C and +3.0°C in the South and in the east coast, while in up-country the difference can reach up to +5.5°C in the Midwest and something around +4.0°C in the North and in the most part of the Northeast and Southeast regions.

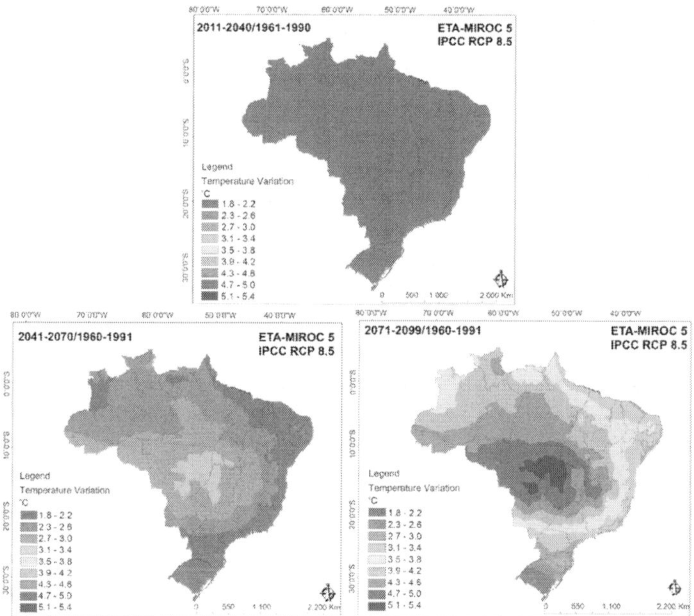

Figure 1. Air temperature differences between the future climate (2011-2040, 2041-2070 and 2071-2100) and the present climate (1961-1990) projected to Brazil through Eta-Miroc 5 regional climate model in the IPCCRCP 8.5 scenario.

In statistical terms, the analyzes indicate a significant upward trend in the average air temperature at 2 m for the five Brazilian regions by the end of the century, confirming the possibility of warming and frequency and intensity growth of temperature extreme events (Table 1). Based on this analysis, the Midwest and North Regions Brazil present the highest rates of average temperature rise, with +1.51°C by slice of time and +1.37°C by slice of time, respectively. The region with the lowest magnitude of increase in the period analyzed is the South, with an increase of approximately +0.87°C by slice of time.

Biofuels (Bioethanol)

The aridity index, formulated in the previous session, is an excellent indicator to assess the water availability in a particular climate type, pointing regions susceptible to the occurrence of severe droughts and desertification process. In the Brazilian, case is notoriously the prone to dryness of the Wild Northeast. From the proposed index (Figure 2), the northern state of Bahia is presented as area of desertification process, which evolves in terms of extension from the time slice 1961-1990 to 2071-2100. It also arise desertification outbreaks in the State of Pará.

The coast of Brazil, both in its east coast as northeast remains with high moisture index, as well as the southeast of the country, which suffers with significant retraction of this area in 2011-2040 period, particularly in the States of São Paulo and Paraná, but from 2041 onwards, it rises again. Certainly the Brazilian coast characteristics associated with a closed topography with high solar radiation and temperature of the surface ensures considerable moisture content and low static stability overlying the air, allowing when the synoptic situation is favorable, the occurrence of convective instability and hence surplus moisture in the environment.

Table 1. Statistical analysis of air temperature at 2 meters for Brazilian regions. Values in red were considered significant at a confidence level of 90%.

Region	Mann-Kendall	Sen's Bend
North	0.09	+1.37°C (by slice)
Northeast	0.09	+1.22°C (by slice)
Midwest	0.09	+1.51°C (by slice)
Southeast	0.09	+1.27°C (by slice)
South	0.09	+0.87°C (by slice)

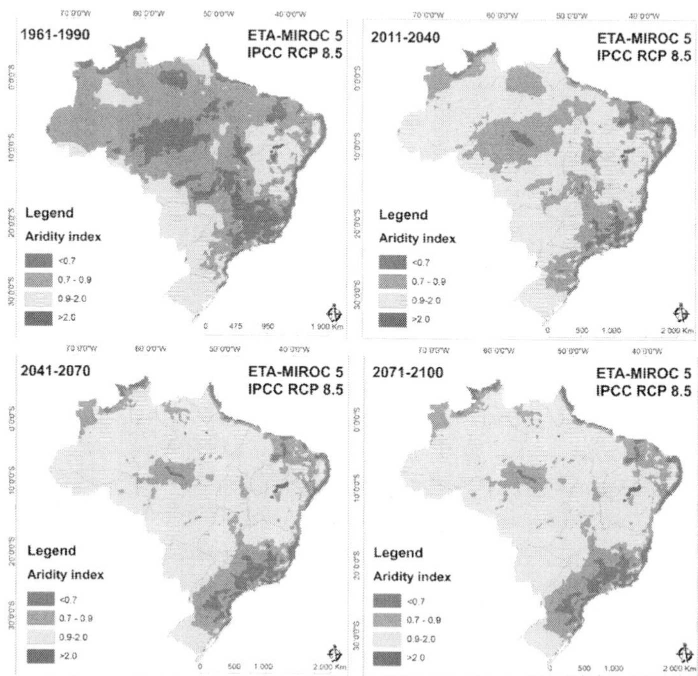

Figure 2. Present climate (1961-1990) and simulated future climate (2011-2040, 2041-2070 and 2071-2100) projected for the aridity index in Brazil through Eta-Miroc5 regional climate model in the IPCCRCP 8.5 scenario.

The most significant impact will be felt in the Central West and North of Brazil, where there is significant reduction in available moisture over the time slices, especially in the Amazon and Tocantins, Mato Grosso and Goiás, with increased susceptibility to dryness situations.

In quantitative terms (Figure 3), the projection data of the IPCC RCP 8.5 scenario from Eta-Miroc5 show that in the period 1961-1990, over 74% of Brazil had aridity index below 0.9, so moisture supply code. However, modeling shows a reversal of this scenario by 2100, where approximately 67% will be in the next dryness class, that is, between the indexes 0.9 and 2.0. The areas in desertification, north of Bahia and in specific situations of Pará, will suffer a 0.2% increase in area extension until the end of the twenty-first century.

As for the statistical analysis, the Mann-Kendall test indicates a significant trend in the level of 90% of the aridity index for Brazilian regions, except for the Southeast (Table 2). The magnitude of this change will be greater in the North of Brazil, which several studies have pointed to the savannization trend of lush rainforest, which corroborates the results. Jarring the other regions, the South has an increased trend of humidity, that is, over the slices, the trend observed in this region is greater availability of moisture.

Soil moisture data (Figure 4) also confirm the spatial trend in the aridity index, so point to an expansion of the water deficit area, especially in the States of Bahia, Piauí and Minas Gerais. The data also show an available surface moisture reduction for the Midwest Region of Brazil and project a little significant change to the Brazilian coast and the Southeast. However, the State of Mato Grosso do Suland Amazon region have significant loss of soil surface moisture over the time slices, common characteristics by virtue of the types of soils developed in these areas.

These data are of great importance, since they are indicative of water availability to the root system of the plant, which has a direct impact on nutrient absorption capacity of it. Thus, from the modeling data is possible to estimate a reduction of the area with water availability for vegetation development, leading directly to the productivity of various genres that are distributed between the Northeast and Midwest regions of Brazil.

Statistical analyzes show a high significance for the Northeast region of the country, whose magnitude is −0.01% reduction in moisture between the slices of time on study (Table 3). The Region that clashes of this feature is the South of Brazil, whose tendency is soil moisture increases of +0.01% between the slices of time, strong by connecting with the results presented to the aridity index in the same area.

Table 2. Statistical analysis of the aridity index for Brazilian regions. Values in red were considered significant at a confidence level of 90%.

Region	**Mann-Kendall**	**Sen's Bend**
North	0.09	+0.04 (by slice)
Northeast	0.09	+0.02 (by slice)
Midwest	0.09	+0.03 (by slice)
Southeast	0.73	+0.01 (by slice)
South	0.09	−0.03 (by slice)

Table 3. Statistical analysis of the soil moisture for Brazilian regions. Values in red were considered significant at a confidence level of 90%.

Region	Mann-Kendall	Sen's Bend
North	0.31	−0.01% (by slice)
Northeast	0.09	−0.01% (by slice)
Midwest	0.31	−0.01% (by slice)
Southeast	0.31	−0.01% (by slice)
South	0.31	+0.01% (by slice)

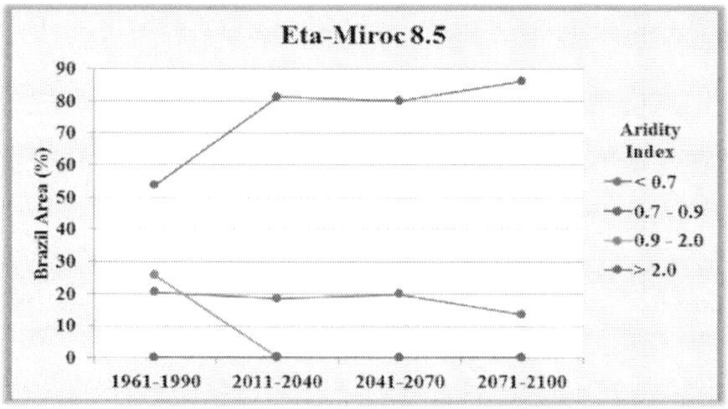

Figure 3. Evolution of the areasofaridity index for Brazil through Eta-Miroc5 regional climate model in the IPCCRCP 8.5 scenario.

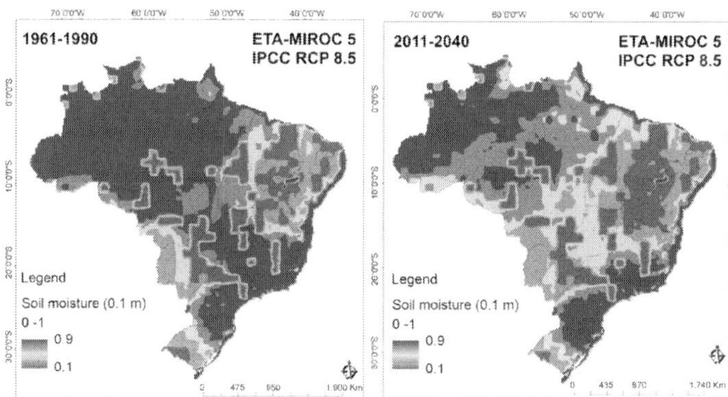

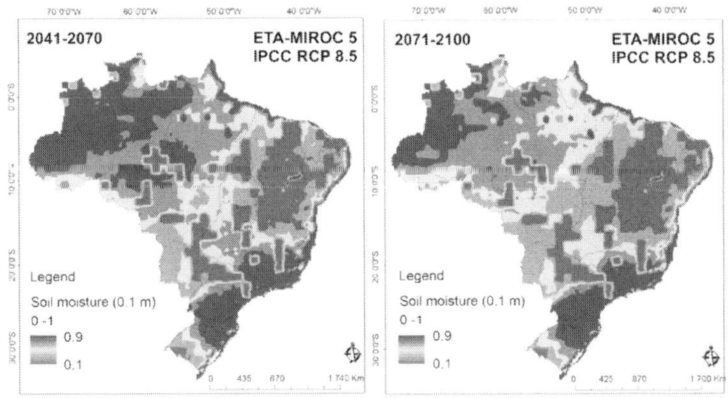

Figure 4. Present climate (1961-1990) and simulated future climate (2011-2040, 2041-2070 and 2071-2100) projected forsoil moisture (0.1 m) in Brazil through Eta-Miroc5 regional climate model in the IPCCRCP 8.5 scenario.

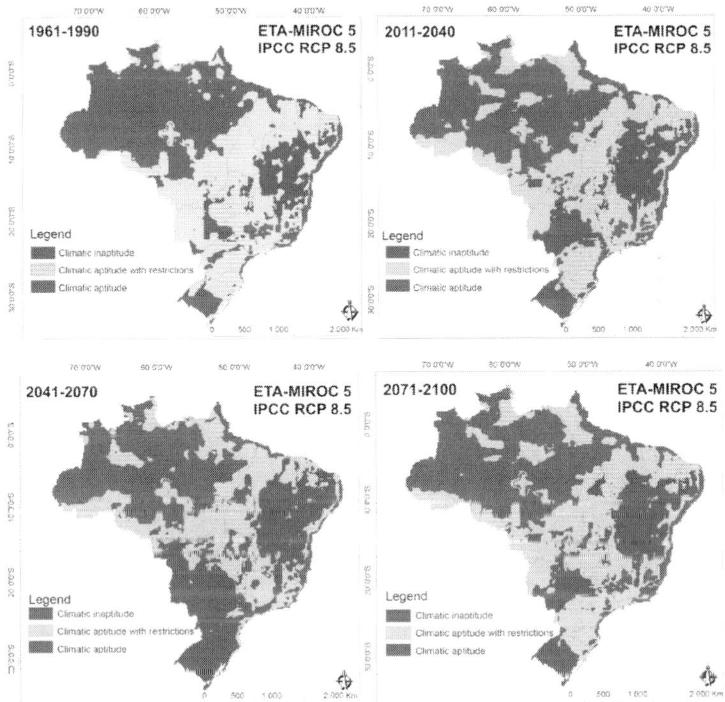

Figure 5. Synthesis map of cultivation fitness class to Brazil for present climate (1961-1990) and simulated future climate (2011-2040, 2041-2070 and 2071-2100) through Eta-Miroc5 regional climate model in the IPCCRCP 8.5 scenario.

From the indicators analyzed above, as well as the air temperature (item 3.1), it was possible the development of Figure 5, which is a synthesis map that shows those areas of Brazil with unfitness for cultivation due to projected climate features (areas in red) as well as those areas which have the potential production, but human interventions with the use of more extensive methods of irrigation are required (yellow areas). The map shows also possible situations of ideal suitability for farming, areas where productivity will reach the maximum possible if the other agronomic and environmental conditions prove favorable (green areas). It is noted that by the end of the century, much of the Northeast of Brazil and in the States of Mato Grosso do Sul and Rio Grande do Sul may present climate unsuited to cultivation. However, most of the coast and the North of the country may be able to farming.

Therefore, from the results presented in this section, it is observed that the Southeast region of Brazil, as well as the country's coastline, is favorable areas for agricultural productivity. However, these regions are historically intended for occupation and development of the urbanization process, that is, with limited land availability and supply for production.

Solar Energy

Analyzing by generalized way all the maps produced by the Eta-Miroc5 regional climate model (Figure 6), there is the same pattern that remains admittedly: an inverse relationship between the intensity of solar radiation and latitude; so at lower latitudes (areas close to the equator, as the North and Northeast of Brazil) solar radiation is high, while in higher latitudes (towards the south of the country), the intensity of this radiation is reduced. This occurs because of the sunlight on the earth's surface, that is, perpendicular in the equatorial and tropical zone and more obliquely toward the poles.

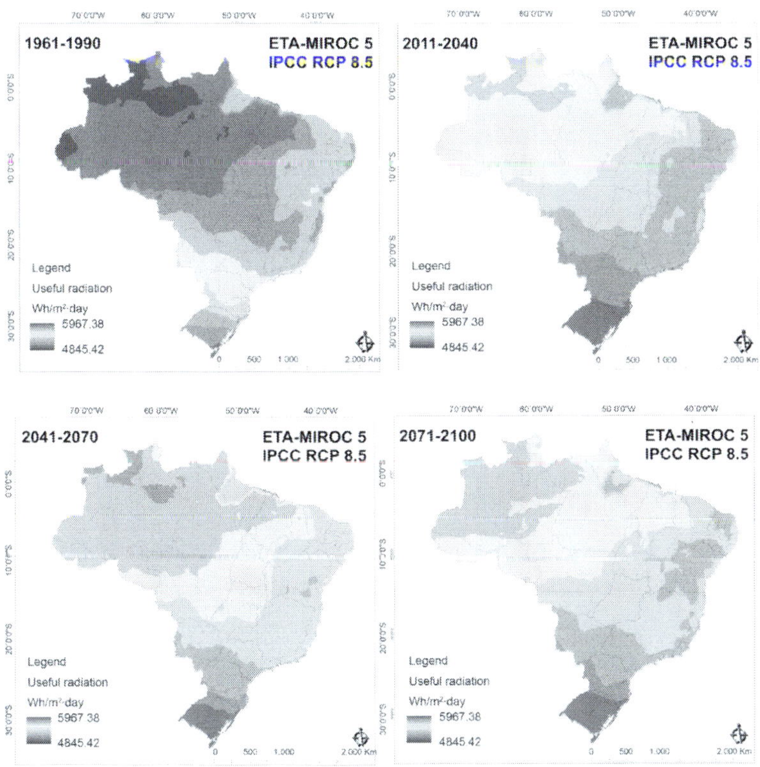

Figure 6. Present climate (1961-1990) and simulated future climate (2011-2040, 2041-2070 and 2071-2100) projected for the useful radiation indicator in Brazil through Eta-Miroc5 regional climate model in the IPCCRCP 8.5 scenario.

When the historical period (1961-1990) of the Eta-Miroc5 model is evaluated, it is noted that the Northern region of Brazil has the highest values of useful radiation indicator, and these are established between about 5700 and 5900 W·h/m^2·day. Next, there are the Northeast, Midwest and Southeast, with values around 5500 - 5700 W·h/m^2·day. The southern region has the most marked variation of intensity of this radiation, with values ranging between 5500 W·h/m^2·day in the north area to 4900 W·h/m^2·day in southern part.

As mentioned previously, the eminent values of useful radiation indicator in northern Brazil are due to emphatic solar energy that this area receives during the whole year. Thus, despite the yearly averages for the indicator representation, the values of them remain significantly high. In contrast, the southern region exhibits lower values due to the lower amount of solar energy receipted by this area. In addition, the South has the striking feature of quite often the performance of weather systems that enhance cloud cover, especially the cold fronts.

Observing the future projections of useful radiation indicator generated by the Eta-Miroc5 model forced with the IPCC RCP 8.5 scenario, it appears that the simulated values by the end of the twenty-first century are minor compared with the present climate (historical period). On average, such radiation intensity reductions vary between 4% and 8% in Brazil, values considered not significant, since the country still find within a suitable range for the generation of energy from the short-wave radiation (solar radiation).

Between 2011 and 2040, the North and the northern sector of the Midwest Brazil must have an average value of the useful radiation indicator in about 5600 $W{\cdot}h/m^2{\cdot}day$, so a decrease of around 4% compared with the present climate. The south-center of the Midwest Region, as well as the Northeast and Southeast regions are likely to have values around 5300 $W{\cdot}h/m^2{\cdot}day$, representing thus a reduction about 6% over the historical period. The South should show greater decay in the intensity of solar energy, about 8%, as the average value in this area would be around 5100 $W{\cdot}h/m^2{\cdot}day$.

Accompanying the future projections over 2041 to 2070 periods, there is a negligible increase in the values of useful radiation indicator in Brazil. These elevations are around 2% compared to the previous period (2011-2040), but still remain around 4% lower in relation to the present climate (1961-1990). In the period 2071-2100, the simulated scenario of solar radiation in the country is quite similar to the first discussed future period (2011-2040), with distribution of values of similar manner to that period. There is only one short exception in the States of Minas Gerais and Paraná, which keeps the value of the previous period (2041-2070).

Figure 7 shows the percentage fractions of Brazil area for certain ranges of useful radiation indicator in function of time simulated by the Eta-Miroc5 regional climate model. Taking into consideration a land area of the country at 8.9 $million{\cdot}km^2$, currently (1961-1990) and future (2011-2100) the majority (around 61%) has/will have an average annual radiation intensity between 5600 and 6000 $W{\cdot}h/m^2{\cdot}day$. On a slightly lower power range (5200 - 5600 $W{\cdot}h/m^2{\cdot}day$) it is noted and projected to remain around 34% of the total area of Brazil. Finally, about 5% of the country's land have/will have solar radiation between 4800 and 5200 $W{\cdot}h/m^2{\cdot}day$.

Also according to Figure 6 and Figure 7, it is noted that the range of greater intensity of useful radiation indicator (5600 - 6000 $W{\cdot}h/m^2{\cdot}day$) is located on the North and parts of the Northeast, Midwest and Southeast of Brazil. On the other hand, the range with lower solar radiation (4800 - 5200 W·h/m²·day) is confined in a given stretch of southern Brazil. This reinforces one of the main geographical features and hence climate of Brazil: a country predominantly tropical with strong solar energy available.

In addition, it is projected a significant increase in average air temperature by the end of XXI century across the country, as seen in item 3.1. It means a reduction in energy conversion capacity, sensitive to ambient temperature variations, especially in the North and Midwest of the country. Furthermore, there would be area limitations for the installation of photovoltaic stations due to the large expansion of the urban phenomenon in the national territory, leaving few free areas for these developments.

In statistical terms, the Mann-Kendal test and Sen's Bend point to a few significant trends to useful radiation indicator in all regions of Brazil (0.73), according to Table 4. Probably, this is due a few factors, including: the use of the 30-year average and the low frequency of variation in solar radiation data. However, from the Sen's Bend analysis, it is noted the magnitude of the trend, with unanimity in whimsy decrease in the useful radiation availability, especially in the North and Midwest, which associated with a rise in temperature could interfere in the production ability of energy from photovoltaic systems. However, it is noteworthy that these values are not statistically significant at the 90% level.

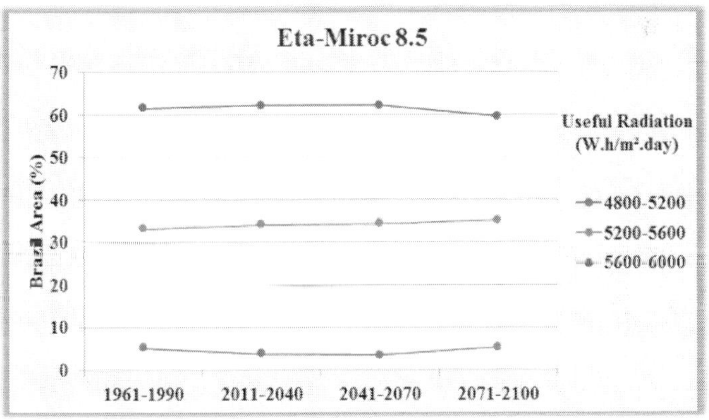

Figure 7. Percentage portions of Brazil area to the useful radiation indicator ranges (W·h/m²·day) between 1961 and 2100 simulated by Eta-Miroc5 regional climate model in the IPCCRCP 8.5 scenario.

Table 4. Statistical analysis of the useful radiation indicator for Brazilian regions. Values in red were considered significant at a confidence level of 90%.

Region	Mann-Kendall	Sen's Bend
North	0.73	−99.56 W·h/m²·day (by slice)
Northeast	0.73	−84.60 W·h/m²·day (by slice)
Midwest	0.73	−92.91 W·h/m²·day (by slice)
Southeast	0.73	−89.57 W·h/m²·day (by slice)
South	0.73	−71.72 W·h/m²·day (by slice)

Wherefore, it can be concluded that Brazil has area and highly expressive power for using solar energy, both today and in the coming decades, especially in the North, Northeast and Midwest regions. Regarding these considerations, it is worth noting that, since the given values are annual averages in 30 years, it is expected that in the summer the incident solar radiation intensity is more significant, with attenuation only in the presence of cloudiness, mainly in the North, a factor not impeding in a tropical country.

CONCLUSIONS

The objective of this study was to calculate indicators and indexes so that they subsidized the analysis of vulnerability and adaptation of the renewable energy sector to climate change in Brazil, focusing this research specifically in the sources of biofuels and solar energy.

For biofuels, in general, the Brazilian coast is presented as propitious area for agricultural productivity, but these are areas historically intended for occupation and development of the urbanization process, that is, with limited land availability and supply for primary production. In some parts of the Northeast, Midwest and South of the country, offer for the cultivation land will be reduced.

In general, for the solar energy, it was observed that Brazil had area and highly expressive power for the use of this power, both today and in the coming decades, especially in the North, Northeast and Midwest. In statistical terms, the Mann-Kendal test and Sen's Bend point to a very weak tendency to useful radiation indicator in all regions of Brazil. In addition, it is projected a significant increase in mean air temperature by the end of XXI century across the country that can mean a reduction in power conversion capability, which is sensitive to ambient temperature variations, especially in the Midwest and North of the country.

Finally, it should be noted that this work is based on current technology, thus technology improvement with biotechnology incentives can reverse the results once presented, and these are the recommendations of this study: 1) investment in biotechnology and genetic improvement of plants for better adaptation to climatic contrasts; 2) investment in technology to better use of solar energy for power generation, in addition to the improvement of the materials to be less susceptible to temperature variation and transmission systems, which can reduce energy losses in the distribution process.

ACKNOWLEDGEMENTS

The authors acknowledge the support of the Brazilian Ministry of Science, Technology and Innovation within the "Preparation of the Third National Communication of Brazil to the Framework Convention of the United Nations about Climate Change" project. The authors also thank Rede CLIMA and Sub-Rede Energias Renováveis for the assistance.

REFERENCES

1. Intergovernmental Panel on Climate Change—IPCC: Climate Change (2013) The Physical Science Basis. Contribution of Working Group I to the Fifth Assessment Report of the IPCC. In: Stocker, T.F., Qin, D., Plattner, G.-K., Tignor, M., Allen, S.K., Boschung, J., Nauels, A., Xia, Y., Bex, V. and Midgley, P.M., Eds., Chapter: Summary for Policymakers, Cambridge University Press, Cambridge, United Kingdom and New York, 1535 p.
2. National Electric Energy Agency—ANEEL (2008) Atlas de Energia Elétrica do Brasil. Agência Nacional de Energia Elétrica, Brazil. http://www.aneel.gov.br/arquivos/PDF/atlas3ed.pdf.
3. Gordon, C., Cooper, C., Senior, C.A., et al. (2000) Simulation of SST, Sea Ice Extents and Ocean Heat Transport in a Version of the Hadley Centre Coupled Model without Flux Adjustments. Climate Dynamics, 16, 147-168. http://dx.doi.org/10.1007/s003820050010
4. Mesinger, F., Chou, S.C., Gomes, J.L., et al. (2012) An Upgraded Version of the Eta Model. Meteorology and Atmospheric Physics, 116, 63-79. http://dx.doi.org/10.1007/s00703-012-0182-z
5. Chou, S., Lyra, A., Mourão, C., et al. (2014) Evaluation of the Eta Simulations Nested in Three Global Climate Models. American Journal of Climate Change, 3, 438-454. http://dx.doi.org/10.4236/ajcc.2014.35039
6. Chou, S., Lyra, A., Mourão, C., et al. (2014) Assessment of Climate Change over South America under RCP 4.5 and 8.5 Downscaling Scenarios. American Journal of Climate Change, 3, 512-527. http://dx.doi.org/10.4236/ajcc.2014.35043
7. Hare, F.K. (1983) Climate and Desertification: A Revised Analysis. World Climate Applications Programme. World Meteorological Organization, Geneva, 149 p.
8. Klein, S.A. (1977) Calculation of Monthly Average Insolation on Titled Surfaces. Solar Energy, 19, 325-329. http://dx.doi.org/10.1016/0038-092X(77)90001-9
9. Sneyers, R. (1975) Sur L'analyse Statistique des Series Dóbservations. Organisation Méteorologique Mondial, Genève, 192 p.
10. Goossens, C. and Berger, A. (1986) Annual and Seasonal Climatic Variations over the Northern Hemisphere and Europe during the Last Century. Annales Geophysicae, 4, 385-400.

11. Sen, P.K. (1968) Estimates of the Regression Coefficient based on Kendall's Tau. Journal of the American Statistical Association, 63, 1379-1389. http://dx.doi.org/10.1080/01621459.1968.10480934

INDEX